TEACHING MATH ONLINE

Great Ways to Differentiate Instruction and Support Parents, K–8

Also by Marian Small

Good Questions:
Great Ways to Differentiate
Mathematics Instruction in the
Standards-Based Classroom
(4th edition)

Math That Matters:
Targeted Assessment and
Feedback for Grades 3–8

Fun and Fundamental
Math for Young Children:
Building a Strong Foundation
in PreK–Grade 2

Teaching Mathematical Thinking:
Tasks and Questions to Strengthen
Practices and Processes

Building Proportional Reasoning
Across Grades and
Math Strands, K–8

Uncomplicating Algebra
to Meet Common Core
Standards in Math, K–8

Uncomplicating Fractions
to Meet Common Core
Standards in Math, K–7

Eyes on Math:
A Visual Approach to
Teaching Math Concepts
(Illustrations by Amy Lin)

TEACHING MATH ONLINE

Great Ways to Differentiate Instruction and Support Parents, K–8

MARIAN SMALL

TEACHERS COLLEGE PRESS

TEACHERS COLLEGE | COLUMBIA UNIVERSITY
NEW YORK AND LONDON

NELSON
nelson.com

▶ Videos, video scripts, and templates for manipulatives included in
 Appendices A and B are available at tcpress.com/teachingmathonline

Published by Teachers College Press, 1234 Amsterdam Avenue, New York, NY 10027;
distributed in Canada by Nelson Education, 1120 Birchmount Road, Toronto, ON,
Canada M1K 5G4.

Text Design: Lynne Frost
Images: Cartoons and photo (p. 42), Shutterstock.com;
  line art and photos (pp. 19, 43, 49), Lynne Frost
Cover Photo: South_agency / iStock by Getty Images

Library of Congress Cataloging-in-Publication Data

Names: Small, Marian.
Title: Teaching math online : great ways to differentiate instruction and support
  parents, K–8 / Marian Small.
Description: New York, NY : Teachers College Press, 2020 | Includes bibliographical
  references.
Identifiers: LCCN 2020034680 (print) | LCCN 2020034681 (ebook) |
  ISBN 9780807764909 (paperback) | ISBN 9780807764916 (hardcover) |
  ISBN 9780807779347 (ebook)
Subjects: LCSH: Mathematics—Web-based instruction. | Mathematics—Study and
  teaching (Elementary) | Mathematics—Study and teaching (Middle school)
Classification: LCC QA20.W43 S63 2020 (print) | LCC QA20.W43 (ebook) |
  DDC 372.7/044—dc23
LC record available at https://lccn.loc.gov/2020034680
LC ebook record available at https://lccn.loc.gov/2020034681

ISBN 978-0-8077-6490-9 (paper)
ISBN 978-0-8077-6491-6 (hardcover)
ISBN 978-0-8077-7934-7 (ebook)

Printed on acid-free paper
Manufactured in the United States of America

Contents

Introduction

SINCE the spring of 2020, teachers have been faced with a new reality, where their interaction with students is not always in the classroom but at a distance. The kinds of normal conversations teachers and students have in the classroom are not possible, students getting answers immediately is not as easy, and students being willing and/or able to ask lots of clarification questions is less likely. Yet teachers still wish to provide rich instruction to maximize the benefit from the limited time during which their students may engage in math learning and to do it in a way that does not overwhelm the parents who are supporting students in this environment. This is particularly challenging for younger students, Grades K–5, for whom parents often have a more critical role, but often even for students in Grades 6–8.

In recent years, researchers have been increasingly exploring the application of technology in education. Although much of the research is focused on higher grade levels, some findings are pertinent at elementary levels as well. For example, Puentedura (2012) offered the SAMR model (**S**ubstitution, **A**ugmentation, **M**odification, **R**edefinition) as a framework for examining different ways that technology can be incorporated into the educational environment. While Modification and Redefinition could be valuable in elementary mathematics teaching, many teachers are focused, at this very difficult time, primarily on Substitution and Augmentation; they are not as interested in exploring the potential of the technology as in moving effective teaching into an online environment, keeping students engaged and learning.

The online environment presents particular challenges in engaging students and in finding new ways to judge student understanding without as much personal contact. Chapter 1 of this resource describes these issues and offers strategies for achieving the best virtual learning environment possible, including helping parents to partner in supporting student learning. Chapter 2 describes the benefits of open questions and parallel tasks—two strategies optimally suited for differentiating instruction, whether in the classroom or online, to reach all students—and highlights challenges and unexpected benefits of working online. Logistical issues unique to operating online at the K–8 level are the focus of Chapter 3, which also describes how to translate classroom tools to online and home environments. The next two chapters present a series of open questions and parallel tasks, illustrating how these strategies can function in the online environment. Chapter 4 focuses on adapting existing tasks for online presentation, and Chapter 5 illustrates creation of tasks particularly suited to the home environment, an important online opportunity to show students how math applies to their everyday lives, a benefit that can shift students' mindsets about math forever and enable them to transfer academic learning to real-life application. A brief conclusion follows (Chapter 6), along with a Resources section that includes not only literature citations but also a list of online sources that provide a wide array of background information, online lessons, and interactive tools of value to teachers, parents, and students.

Two valuable additional resources are provided with this volume. I have created eight videos as models of videos that can be sent to parents; these may be accessed from the Teachers College Press webpage for this book (tcpress.com/teachingmath online). Five of the videos focus on applying important math tools in the online environment, and three illustrate the use of probing questions in stimulating rich math conversation. Scripts for these eight videos are included in Appendix A and at tcpress.com/teachingmathonline for teachers who would like to use the scripts as bases to create their own videos or online lessons. Finally, Appendix B includes templates for common math tools that can be used in the home. These templates, and others, are also available online (tcpress.com/teachingmathonline).

▶ **VIDEOS & MATH TOOLS:**   Available at tcpress.com/teachingmathonline

---

MY INTENTION with this resource is to show how materials teachers already have and use on a regular basis might be appropriately adapted to help enrich mathematics instruction in the virtual environment, and to uncover new ways to create home-based learning and partner with parents. I hope readers will find it useful.

CHAPTER 1

Principles to Consider When Teaching Online

IN THE CLASSROOM, teachers use many strategies to engage students, to pick up on nonverbal cues when there appears to be confusion, to encourage attention, to encourage responsiveness, and to build a warm and risk-free environment. In a virtual environment, the inherent distance between students and between teachers and their students makes this more difficult, but no less important. A number of the principles I discuss in this chapter to achieve the best virtual learning environment possible are adapted from Darby (2019). More detail about ways to achieve these goals is provided in subsequent chapters. But, before talking about these principles, it might be valuable to first emphasize that what matters most is our beliefs about the teaching of mathematics that underlie instruction, whether in the classroom or virtually.

WHAT MATH TEACHING MUST FOCUS ON

For some time, a broad community of voices has been stressing the importance of teaching mathematics equitably to ensure that all students benefit from rich instruction (Huinker et al., 2020). My commitment to this goal has led me to create numerous resources designed to (1) make sense of math for students, focusing on conceptual understanding as the goal of instruction (e.g., Small, 2017, 2019), and (2) make math accessible and rich for *all* students, from those who are struggling to those who are performing at a high level (e.g., Small, 2020). You might consult any of the listed resources to learn more about what sort of math I believe we must focus on in our teaching. Those beliefs will be implicit in the examples used in this resource, but the focus here is less on how the tasks are developed and more on how such rich teaching can be nuanced for an online environment.

CREATING WARMTH

Students respond to the warmth and positivity of their teachers in a classroom setting. Therefore, teachers must find alternate ways to convey warmth and positivity at a distance. These include the following:

- physical energy displayed in online conversations
- positivity displayed in how tasks are provided as well as in how feedback is offered.

Teachers must be smiling, engaged, and energetic in their online interactions with students. They need to present tasks in a manner that signals that they have confidence in their students' ability to handle the tasks, for example, *"I found this task I just know you're going to love."* or *"I know you are going to do such awesome work with these problems."* As well, feedback can point out all of the things about a student's work that are interesting, correct, and so forth, even when challenging students to reconsider or extend ideas.

CREATING INTERACTIVITY AND ENGAGEMENT

The best classrooms, whether in mathematics or in some other subject, are ones where there is lots of conversation, attention to and building from others' ideas, and tasks and tools that foster engagement and interactivity. Students often talk with each other and not just the teacher or the whole class. Again, it is important to replicate as much of this as possible in a virtual environment.

Platforms that are used in virtual environments are normally those supported by the particular school district in which a teacher works, but teachers should advocate for platforms that make communication between students and between students and teachers easiest. Especially for young students, the mechanism for communication needs to be intuitive and easy. Platforms that allow for break-out discussion groups are particularly useful for older students.

Sometimes interactivity should be in small groups, so, whether using the main teaching platform or not, teachers should seek ways to create pairs or small groups of students, which might change regularly, who are expected to interact to work on tasks together, especially for students beyond the earliest primary years. Teachers might suggest ways to communicate, such as Google Meet or video chat on a phone or simply talking by phone. A teacher might also suggest times when students explain their thinking to another student, rather than the teacher; the peer can then be expected to comment to the teacher about what he or she learned from the buddy's explanation.

It is valuable to have opportunities to show student work to other students (with permission) so students can learn from each other as much as they would in a classroom where a teacher would display work. There can be digital galleries of work.

Engagement often increases when tasks are presented in a more visually appealing manner. This might be considered when tasks are initially presented.

Engagement also increases when teachers are responsive. It is important to respond to students' comments in such a way that it is very clear to students that their thoughts were heard. This might involve rephrasing or asking a specific follow-up question that uses the child's words.

Just as there are behavior norms in the onsite classroom, there must be behavior rules that are discussed with and clarified for students in the online environment. These rules should make sense to students, and students might be invited to help suggest some behavior norms. The consequences for not abiding by the norms must be clear, and there must be follow-through.

Engagement can be monitored in a number of ways, whether it is simply a student's presence online or whether students respond to most tasks assigned. Older students might be expected to maintain a log online recording how long they spent on various tasks, how long they talked with a classmate about a task, or how often they attempted to communicate with the teacher without prompting; parents of younger students might provide some rough estimates of time spent on assigned tasks.

PROVIDING CLARITY AND STRUCTURE

There are different aspects of online learning that require clarity and structure.

For young students, providing their parents with an overview of what to expect in a given week is probably helpful in allowing the parents to support their children. A schedule could go to older children directly. As well, information about how and when a teacher can be accessed is also important for both parents of younger students and for older students to receive. Students and parents should also understand the various reasons they might contact a teacher: for clarification of a task, to get help, or to request more challenging work, for example.

Make sure that navigation tools within the platform are easy to use and clearly explained both to students and, for younger students, to their parents.

To ensure that tasks are clear, it might be valuable for the teacher to read through the tasks beforehand several times, trying to see them as a parent or student would, making sure that they will understand exactly what is being asked.

Although a teacher might be tempted to provide lots of examples, this is actually not desirable when a teacher is seeking divergence and not convergence, as is the case with open questions. Some scaffolding, but not too much, is ideal. Examples are described in Chapter 4.

In some jurisdictions, teachers are being told how long to be online with their full classes. If teachers have some latitude, for younger students, online sessions work best if they are short, perhaps 20 minutes maximum, but for older students, an hour might work, though shorter periods are probably preferable. Teachers might have instructional sessions three times a week, but be available the other two days, while students are working asynchronously, to answer questions, check in with a few students, and so on.

When students receive a task, it needs to be clear whether student work needs to be returned or not or whether the task is simply a learning opportunity that will not be monitored by the teacher. If work is to be returned, the deadline needs to be clear, as do expectations for whether the response is meant to be oral or written or either, as well as what pieces are required in a submission.

There is value in arranging some "social" situations where all students in a class participate in a virtual meeting to maintain the class a unit, but most instruction is probably best accomplished with smaller groups. This is to ensure that the instruction does not become a "lecture" and that students really can interact with the teacher—and that the teacher can handle the interaction.

ATTENDING TO INDIVIDUAL DIFFERENCES

In an onsite classroom, there is always accommodation for students with special needs, whether it is a physical issue or a language issue. Similar accommodations need to be considered in an online environment. These might be providing text that is read to students or text that is easy to enlarge or text that has been translated into another language. Teachers might use language translation sites or apps, such as Google Translate, to translate tasks into students' home languages, and they might allow for student responses to be in their first language.

In fact, an online environment might make it easier to attend to student differences than does the classroom. A teacher could create small groups within the class for different purposes and link to those small groups in different ways or provide support differently than for other students.

Teachers may notice that students who are often quiet in the classroom are quite comfortable speaking online, or vice versa.

◈ CHAPTER 2 ◈

Differentiating Instruction Using Open Questions and Parallel Tasks

TEACHERS might assume that it is easier, in a completely virtual environment, simply to locate and assign online worksheets or short-answer questions for students to complete to cover the various topics they need to teach, but there are a number of problems with this approach:

- These types of tasks might reveal whether a student can perform particular procedures, but rarely do they help students make sense of new ideas.
- If these types of tasks are used, the teacher cannot know who actually did the work. Was it really the student, or was there so much "help" that the student really did very little?
- If these types of tasks are used, can the teacher know what the student was thinking as he or she determined those answers?

So, let's go back to why we might use open questions and parallel tasks in the classroom and see how those principles apply in an online environment.

THE VALUE OF OPEN QUESTIONS IN AN ONLINE ENVIRONMENT

An open question is one for which a variety of quite different responses might be possible and correct. For example, instead of asking the more closed question *What is 8 × 9?*, a more open question could be any of these:

- *The answer to a multiplication is more than 50. What might you have multiplied?*
- *You multiply two consecutive numbers. The ones digit of the product is 2. What could you have multiplied?*
- *Can you multiply two numbers and get a result about 60 more than one of them?*

Notice that there are many answers to the first question, for example, 52 × 1 or 9 × 6 or 8 × 9, or many others. To get those answers, students access many mathematical ideas, for example, multiplying by 1 results in the number you were multiplying; if one number is less than 5, the other must be more than 10; and so forth. The question also is likely to lead to a great deal of multiplication practice.

Similarly, the second question has many correct responses. Some examples are 8×9 or 1×2 or 18×19 or 3×4 or 23×24, and so on. This is likely to lead students to consider how only the units digits matter in determining the units digit of a product when multiplying multidigit numbers and encourages students to go through their multiplication facts.

There are also many responses to the third question. These include, for example, 8×9 (since 72 is about 60 more than 9) or 61×1 (since 61 is 60 more than 1) or 31×3 (since 93 is about 60 more than 31). This question, too, encourages thinking about how multiplication works and provides multiplication practice.

In a regular classroom setting, the use of open questions is valuable for several reasons:

- It allows for differentiation of instruction, so that students at different spots on a learning trajectory can still respond and benefit from the elicited thinking, but at their own level. (Many years ago, Vygotsky [1978] focused on the need to provide tasks within each student's *zone of proximal development* to ensure that each has a true opportunity to learn and each can make a meaningful contribution to the class community. More about research on the importance of differentiation can be found in Small [2020].)
- It allows for richer interactions when students get to hear each other's responses.
- It ensures that the focus of instruction is more on the underlying important ideas and less on the details of posed questions.

Look, for example, at this question that a teacher might ask in a Grade 3 mathematics classroom:

> Two fractions are fairly close to 1, but one is just a little closer than the other. What could the fractions be and how do you know you are right?

One student with less comfort with fractions might choose $\frac{1}{2}$ and $\frac{3}{4}$. But some other students might think these fractions are not sufficiently close to 1 because they envision something like this:

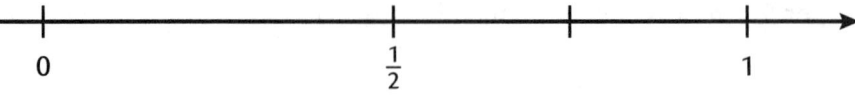

But what if the first person was envisioning this?

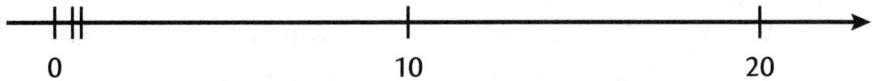

In the second instance, $\frac{1}{2}$ and $\frac{3}{4}$ both seem quite close to 1.

The interaction between students about whether that second number line is "legitimate" leads to a rich math conversation. But other mathematical ideas are likely to come up as well. For example, a student might argue that $\frac{5}{6}$ is close to 1, but $\frac{6}{7}$ is closer. The argument would be that the first fraction is only $\frac{1}{6}$ away from 1 (which is not a lot), but the second is only $\frac{1}{7}$ away (which is closer). This argument helps fellow students learn several important math ideas:

- Closer means a smaller distance between two numbers on a number line.
- $\frac{1}{a} < \frac{1}{b}$ if $a > b$, when a and b are counting numbers.

The focus becomes less on whether the answer is correct, and more on what makes one fraction closer to 1 than another, as well as the effect of context in deciding what *close* means, another important mathematical concept.

THE VALUE OF PARALLEL TASKS IN AN ONLINE ENVIRONMENT

Parallel tasks can also be useful in online mathematics instruction.

The notion of a parallel task is an original task and then a modification offered that either simplifies or makes more complex the details of the task without altering the main idea being explored. Students choose or the teacher assigns the task they will perform, and in that way, students at different levels can engage in important mathematical thinking, at an appropriate level, and learn from students who share their thinking on the other version of the task.

For example, two parallel tasks might be:

> *Choice 1:* On a number line, you begin at 4. You move forward 9, then 10, then 11, and then go back 8, then 9, then 10. Where do you end up?
>
> *Choice 2:* On a number line, you begin at 34. You move forward 19, then 20, then 21, and then go back 18, then 19, then 20. Where do you end up?

The identical idea appears in both tasks, that you've really moved forward just 3, since you can rearrange what you've added and subtracted to make the calculation easier. But if a student simply follows the rules, in order, one calculation is simpler than the other.

When parallel tasks are used in a classroom, the whole point is to ask what one might call *common questions,* that is, the same questions no matter which task is completed. In this way, students learn from each other, with the focus remaining on the important ideas.

In this case, some common questions might be:

- *How could you have predicted you would end up ahead of where you started?*
- *Did you have to do the forward and backward moves in the order given, or does it not matter?*
- *What strategy did you use to decide where you would end up?*

In an online environment, parallel tasks also work well. Students can still be given a choice of which of two tasks they wish to do, and the common questions can be posed and then discussed, ideally in an interactive synchronous situation, so that students can hear and react to responses of other students.

THE VALUE OF RICH MATH CONVERSATION: UNEXPECTED BENEFITS OF AN ONLINE ENVIRONMENT

One of the major strengths of open questions in differentiating instruction is in the way this approach strategically stimulates rich math conversation. At first, it might seem that the online environment—with its physical separation of students and teachers—might inhibit such conversation.

However, the extra time afforded in an online environment gives students a chance to think about their responses to questions and post them, followed by an opportunity to interact with other students about their responses. It would not take a lot of time (which is important in this setting with young students especially), but the discussion would be rich, and all students, from the weakest to the strongest, are likely to benefit, as are the parents who are likely supporting all of this.

As well, since many students are apt to post their responses with short videos, the teacher can gain a better sense about what their students know or are still confused about than would be likely with a series of right/wrong responses on a worksheet or during a fast-moving classroom conversation. And without the peer pressures of the classroom environment, students who are struggling may be more likely to seek additional clarification when needed.

A particular form of conversation that is powerful, even in mathematics, is a debate, where students legitimately hold different opinions on an issue. For example, students could take a position on the issue of whether they think a better way to figure out 53 − 19 is to add up from 19 to 53 or take 19 away from 53. They could then plan their argument, perhaps with another student. Students could communicate their positions to the teacher digitally, and the debate could happen with a group of students online, where the teacher ensures that students on both sides of the debate participate together.

THE IMPORTANCE OF BEING INTENTIONAL IN AN ONLINE ENVIRONMENT

One idea that teachers struggle with, even in a regular classroom setting, is deciding what to focus on when teaching a topic. Should the focus be on how to answer different kinds of questions or should it be on ideas? And if it is ideas, what ideas matter?

Generally, standards do not really tell teachers where to focus, and resources that teachers draw from often take positions on what to focus on, but do not share with the teacher why those positions were taken. This does not allow the teacher using the materials to consider alternate decisions that might have been made.

Given that an online environment with young students normally involves limited interaction time, it becomes increasingly important for teachers to be intentional. For example, if a teacher is trying to teach students about percent, the teacher must decide what ideas about percent to focus on.

Is the most important objective:

- How to calculate $a\%$ of b by multiplying $\frac{a}{100} \times b$ (e.g., 35% of 25 = 0.35 × 25)?
- How to determine a number if $a\%$ of it is b by dividing b by $\frac{a}{100}$ (e.g., 26 is 40% of 65 since 26 ÷ 0.4 = 65)?
- Estimating percents?

Or is it:

- Realizing that if you know $a\%$ of b, you automatically know a lot of other percents (e.g., if you know that 20% of 50 is 10, you also know 20% of 2 × 50 and 40% of 50, etc.)?
- Realizing that solving a percent question is about determining an equivalent ratio (e.g., 30% of 80 means that $\frac{x}{80}$ is equivalent to $\frac{30}{100}$)?
- Realizing that we use percents to standardize ratios to make them easier to compare?

When using open questions, parallel tasks, or more traditional sorts of tasks, being intentional becomes increasingly important in an online environment, given the additional time constraints.

❖ CHAPTER 3 ❖

Logistics

THERE ARE many very specific issues to deal with when teaching elementary aged students in a virtual environment. A number of these are addressed in this chapter.

PREPARING STUDENTS FOR WORKING AT A DISTANCE

It is a good idea to ensure that students understand online expectations. They should know how often they are expected to be online, what sort of behavior is expected in online situations, and what happens if there is bad behavior. They should know how much participation is expected and how to participate.

For offline work, they should be helped to understand what tools they will need, the value of a quiet space for working, and the choices they have for getting help when they are unsure.

Time expectations should be made clear. A teacher might begin any online session with students with an indication of how long the session will last and provide, with any offline task, an indication of how much time is likely to be spent on it.

Students should also be clear about how sharing online will happen. Do students just take the microphone and talk, or is there a signal? If a task is assigned, it must be made clear to students whether their work is expected to be submitted, and how and when it is to be submitted.

It may be valuable to provide students or parents with URLs for websites that familiarize them with the use of online tools, whether using a meeting platform, learning how to submit a short video, or some other aspect of online learning. For the sake of parents with limited English, and for parents who find it easier to learn new technology via video, it would also be important to provide a video describing key tools such as chat, hand raising, creating and posting video, and submitting work using the school's platform. Of course, teachers need to make themselves familiar with these tools first.

CONNECTIVITY ISSUES

It has become clear that connectivity issues might create problems for students either because their home has no internet access or because many people in their home are working online simultaneously, competing for limited bandwidth.

It is not always possible to solve these problems, but a few actions might help. It might be useful to not use gratuitous images (although some images are probably critical). It might also be useful to consider how communication through cell phones might help, being aware, of course, of potential costs for some cell customers. As well, if lessons are recorded, students can look at them at a later time, when bandwidth might be less of a problem.

BUILDING ONLINE COMMUNITIES

One of the most important aspects of school is the social interaction it affords children with each other and with their teacher. It is important to find ways not to lose that in a virtual environment. Most of us are not set up to learn alone.

That said, interacting in an online environment with their teacher and classmates is less comfortable for many students. So, how a teacher builds online communities is critical.

Although for social purposes, the teacher might have online meetings with their whole class, the teacher might create short meetings with a smaller part of the class (perhaps a third or a quarter of the students) for instructional purposes. But even those meetings need to start with some social interaction. For younger children, particularly, limit the direct instruction time to about 10 minutes; even for older students, keep these sessions to not much more than 15 minutes. The teacher might mix up the groupings for instructional purposes so that students get to see all of their classmates over a span of time.

The teacher might set up short meetings for these different purposes:

- Providing background to make an assigned task make sense
- Doing a little bit of math together (e.g., a short number talk)
- Meeting with a small group of students to reteach an idea they are struggling with
- Meeting with students seeking more challenge to provide an extension task suitable for them
- Consolidating some work and drawing out the important ideas after a task has been completed offline

The teacher might consider either encouraging students to pair up or assigning each student a buddy in the class to talk things over with before submitting any of their own work. For older students, a teacher could ensure that working in pairs is not only allowed but encouraged; some students are likely to want to choose their own buddies, although a teacher might suggest combinations instead.

If teachers provide anything in text form for students, they might take advantage of hyperlinks in the documents to provide small clarifications or extensions that students would find helpful. An online environment allows the site or document creator to use hyperlinks, thereby limiting the amount of text students need to get through in one pass.

For example, if teachers were providing a summary of ideas for students on a topic such as how to multiply two 2-digit numbers, they could provide some print instructions that show what the operation looks like. For example:

$$
\begin{array}{r}
^{1}24 \\
\times\,23 \\
\hline
72 \\
480 \\
\hline
552
\end{array}
$$

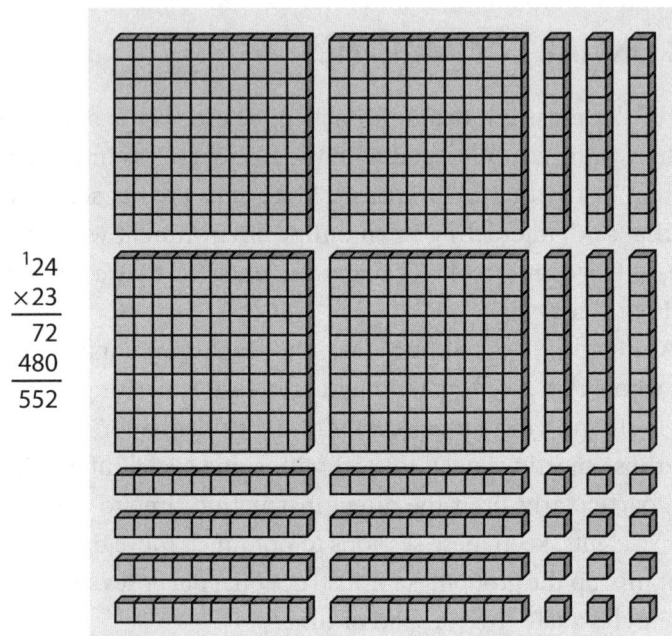

But they could also provide a video to show how the steps are performed, one at a time, talking out what is written, when, and why. (See the description in the section on manipulatives that follows.)

Teachers must identify ways for students to communicate directly with them. They might explain to students the concept of *office hours* and provide office hours for students each day or on alternating days; or, different office hours for shorter amounts of time might be established for different groups of students. Office hours might be used to help students struggling with a task, or for a student to clarify instructions for a task. Office hours might be used for a student who wants an additional challenge if the task that was provided was quite easy for that individual.

As stressed earlier, it is important in the virtual environment to show understanding and warmth and positivity in all interactions with students.

If teachers end up teaching in a hybrid environment, some online and some in person, it might be best to try to accomplish most of the new instruction in the in-person environment, using the virtual environment primarily for follow-up.

COLLECTING WORK FROM STUDENTS

I have observed that teachers often feel remiss if they don't collect an abundance of work from their students. However, a few carefully considered pieces might provide the same or even better information to the teacher without the stress that submitting a lot of material causes for students. Pieces that are thoughtfully constructed can focus on the more important ideas teachers want students to learn without requiring many, many examples of their execution.

Ideally, the pieces I would want students to submit would normally be videos or audio files where students are talking through the task and explaining their thinking rather than providing only solutions.

For example, a teacher might provide three tasks in a given week focusing on subtraction of 2-digit numbers using different strategies. Students then might be asked for only one submission where they pick a subtraction question they answered, describe what strategy made the most sense to them for that question and why, clarify how and why the strategy worked, and also describe one other strategy they could have used.

It might be useful to provide written success criteria so students can ensure they accomplish the entire task. In this case, criteria might be:

- ☐ I choose one of the subtractions I had to do.
- ☐ I make a video where I describe how I did it and why I did it that way.
- ☐ In the video, I also tell how I know I am right.
- ☐ I also tell another way I could have done the question.

Teachers must remind the students, perhaps, that they have to make sure they satisfy each of the success criteria.

A number of easy-to-use online tools allow children to submit videos of themselves talking in either a synchronous or an asynchronous situation. One example is Flipgrid (see the Resources list at the back of this book for where to obtain it), which allows for synchronous discussion of submitted videos, but there are others. Capturing a video on a cell phone is also a possibility.

In other cases, a teacher might take a different approach: Provide success criteria to students for an assigned task and ask students to submit an audio or video file in which they talk about how they met the criteria. Ideally, the criteria should not cover just the steps in getting a correct answer but also should include explanations of the student's thinking.

For example, if a teacher had assigned a task asking Grade 7 students to describe what numbers one could or could not represent by adding a positive integer to the opposite of its triple (e.g., adding 4 to –12 or 20 to –60), the success criteria might be:

☐ I describe lots of integer results I can get and tell why those are possible.

☐ I describe lots of integer results I could not get and tell why those are not possible.

Students could then share their thinking about how they met the criteria.

PROVIDING FEEDBACK AND FORMATIVE ASSESSMENT

As would be the case in the classroom, the focus should be on formative assessment and feedback, rather than summative assessment.

It is important that feedback be focused on important learning goals and not simply counting how many items students got correct or incorrect (Small, 2019). For example, feedback on the subtraction task described in the preceding section might focus on whether students clearly explained why their chosen strategy was preferable, whether they applied the strategy well, and whether they knew what alternative strategies were possible. In addition, they might be asked why a different strategy they learned or that is proposed might not seem as good.

For the integer task described earlier, especially if a student just tried examples rather than using algebra, feedback might focus on whether students had tried enough examples to see that no positive integers were possible and that no odd integers were possible, and on asking students if they could describe all the possible and all the impossible results in the situation, to help them consolidate their thinking.

INVOLVING PARENTS WITHOUT OVERWHELMING THEM

Schooling students at home places an added burden on parents, many of whom are dealing with several children and often their own job requirements. For this reason, even though parents are more than likely to be called upon by younger students no matter what is sent home, it is important to try to minimize the load on those parents.

It seems inappropriate to ask the parent to become a math teacher. It is the teacher's job to provide most of the instruction. Nevertheless, it is valuable to provide parents with guidance through quick and easy-to-read instructions in print form and/or very short videos explaining what the tasks the teacher is assigning are about, how the math works, and why the tasks have been selected.

It is probably helpful to focus on more familiar math topics, although they still must be taught in a rigorous and thoughtful way, as suggested by Huinker et al. (2020). That might mean focusing on some familiar topics that are especially important at a particular grade level and deepening understanding in those areas, while saving instruction for other less familiar (to parents) grade-level content until children are back in school. For example, a Grade 7 teacher might skip over topics dealing with statistical variability or probability simulations in the online environment and focus instead on topics involving proportional relationships, working with rational numbers, algebra, and circle measurements; the other topics would be addressed when students are back in the classroom if that becomes possible. Forcing parents to deal with extremely unfamiliar content probably places too much of a burden on them if it can be avoided. Teachers have to make their own professional judgments about what they see as more or less suitable when parents have a larger role to play; there is no one-size-fits-all, definitive approach. Creating priorities is inevitable (Council of the Great City Schools, 2020).

An equally important consideration is the manipulatives that would be required to make sense of various topics. If those materials are hard to come by in the home (e.g., pan balances), those topics should probably wait for onsite instruction if at all possible. I discuss virtual manipulatives in the next section.

Parents may need a quick meeting or a note from the teacher encouraging them to support their children not by showing the children how to correctly perform the assigned task, but instead by asking probing questions to move the students forward. This will only make sense if the teacher clarifies that the goal in this online environment is to hear student thinking more than to check for correct answers, even though that will also be done. (Numerous examples of probing questions are provided in Chapters 4 and 5, and three videos illustrating the use of probing questions with tasks described in Chapter 4 are available at tcpress.com /teachingmathonline, along with scripts; the scripts are also available in Appendix A.)

▶ **VIDEOS:**   *Using Probing Questions:*   234 Question [Grades K–2]
               Using Probing Questions: What Does Division Look Like? [Grades 3–5]
               Using Probing Questions: Circles, Squares, and Crosses [Grades 6–8]
               Available at tcpress.com/teachingmathonline

Consider taking advantage of students being home and assign activities that really suit exploring math in that environment. In this way, students might begin to appreciate that math is not an abstract pursuit that applies only to the classroom, but instead is part of everyday life. Tasks like these could be a start:

Grades K–2: Counting how many spoons are in the house
Grades 3–5: Asking what fraction of the area of their bedroom is covered by beds or is not covered by furniture
Grades 6–8: Asking them to use percents to describe how full various containers, closets, and so forth, are

Ideas for such activities are provided in Chapter 5.

DEALING WITH MANIPULATIVES WHEN TEACHING VIRTUALLY

Manipulative materials and other visual tools are an essential part of teaching math at the Grades K–5 level, and ideally at the Grades 6–8 level as well. Schools usually have stocks of such manipulatives, but individual homes generally do not. Teachers might consider sending small baggies of materials home by making them available at the school at a particular time for pickup, but this may not be practical. So teachers need to consider alternatives to avoid going back to the days of teaching math only symbolically.

Fortunately, online manipulatives and apps are widely available. To help parents know where to begin, the teacher could provide links or URLs for an assortment of sites where they (or the students) can locate resources that are available free of charge. For example, Utah State University hosts a National Library of Virtual Manipulatives, and other materials and activities are provided by The Math Learning Center, by learning resource companies such as Didax, and by Mathies (sponsored by the Ontario Ministry of Education). One site that gives access to many others and is quite convenient is The Techie Teacher. URLs for these sites are provided in the Resources section at the back of this book.

Sites vary with respect to ease of use of the manipulatives offered. Make sure that the sites provided to students and parents are "intuitive," that is, that it is very easy to figure out how to use the manipulatives. Teachers must work with the sites themselves in advance to check them out and should provide a few simple screen shots for parents to assist them in getting started.

Limit the variety of manipulatives required to make their use less onerous.

Provide, for parents who want them, two-dimensional visuals that they (or students) could cut out for the following manipulatives:

- 10-frames
- 100-chart
- Number paths
- Number lines
- Pattern blocks
- Dot paper
- Addition table
- Base ten blocks
- Tangram puzzle
- Grid paper
- Multiplication table
- Fraction strips
- 10 × 10 grid
- Double number lines

Templates for many of these items are provided in Appendix B of this book; those and others are available for free download from tcpress.com/teachingmath online. The manipulatives could be printed out as needed at home or could be mailed or made available for pickup at the school for homes without printers. Many students, but certainly not all, respond better to tactile manipulatives than to virtual ones.

▶ **MATH TOOLS:**   Templates for Manipulatives
                    Available at tcpress.com/teachingmathonline

---

When assigning tasks requiring the use of counters or linking cubes, edit the tasks to suggest use of items readily available at home, such as buttons or Lego blocks. Egg cartons with two cups cut off could serve as 10-frames.

Take advantage of the online environment to provide annotated videos or narrated PowerPoint presentations for both students and parents on how various manipulatives are used. The videos for students need to be very succinct, but even the videos for parents should be kept brief. Make sure to say something about how the use being modeled for the manipulative makes mathematical ideas easier to see. Be as specific as possible and relate the explanation to an activity that students must complete.

Examples of videos I have created to explain the use of manipulatives are available at tcpress.com/teachingmathonline. Scripts for these videos are included in Appendix A and are available at tcpress.com/teachingmathonline.

▶ **VIDEOS:**   *Learning About Math Tools:* Using a 100-Chart [Grades 1 and 2]
                Learning About Math Tools: Using Number Lines [Grades 2 and 3]
                Learning About Math Tools: Using Pattern Blocks [Grades 3 and 4]
                Learning About Math Tools: Using Base Ten Blocks [Grades 4 and 5]
                Learning About Math Tools: Using Double Number Lines [Grades 6 and 7]

                Available at tcpress.com/teachingmathonline

❖ CHAPTER 4 ❖

Adapting Questions From *Good Questions* for the Online Environment

IN THIS CHAPTER, I have pulled some questions from my book *Good Questions: Great Ways to Differentiate Mathematics Instruction in the Standards-Based Classroom,* 4th edition (Small, 2020), and show how one might adapt them for a virtual teaching environment. Throughout, one objective is to make the activity more visually appealing, if possible, since students are likely to expect more dramatic visuals online. Ideas for attractive graphic representations are included for some of the examples that follow, but not for every one.

GRADES K–2

The answer is 5. What is the question?

This question might be more appealing to a student when presented like this:

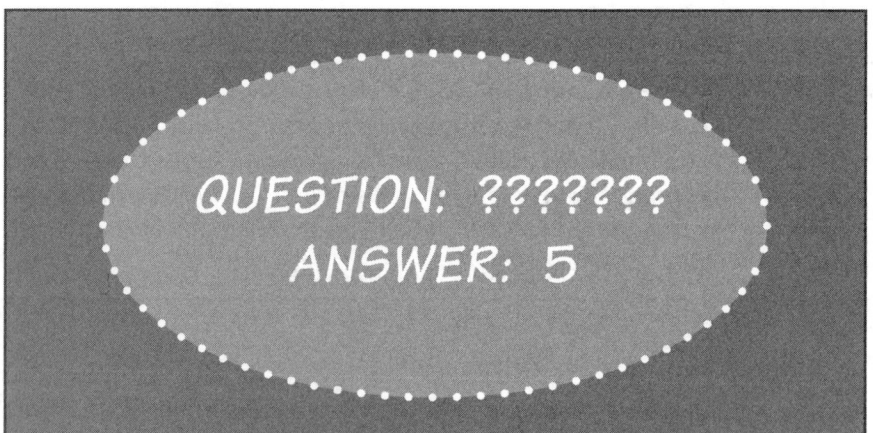

This question could be posed in a virtual environment, but students will likely wonder what sort of answer the teacher is seeking. As presented here, the question is designed for younger students, where responses such as *What is 8 – 3?* or *What is 2 + 3?* are expected, although it could also be posed for older students with more sophisticated answers welcomed (e.g., *What is the square root of 25?*) A teacher might be tempted to provide an example of a correct response with the task, but if that is done, it is highly likely that every response will be a replica of the provided model and, as a result, the question will not foster students' own thinking.

Instead, a teacher might set up a quick online meeting with students where the teacher talks about the difference between a question and an answer. For example, the teacher might share that a question might be *What is your name?* And the answer might be *Sarah.*

Point out that usually teachers ask a question and the students give an answer, but in this case the provided task is backward. This time, the students know the answer but not the question. Make sure students realize that this time the answer is 5, and they have to make up a question where the answer is 5. Encourage them, in fact, to make up a variety of questions.

A teacher should Indicate to students that there are lots of possibilities for questions they might create. Reassure them that there is not a particular question that is the best one, but that a variety of questions will be appreciated.

It might be a good idea to share success criteria. For example:

☐ I give a few questions where the answer is 5.

An additional success criterion could also be included, but need not be:

☐ I tell how I know the answer is 5.

Students go offline to work on the question. Teachers might provide parents with a recording of their meetings with the children so the parents can replay it. They might also provide some suggestions parents can use if their children seek support, but teachers should also make themselves available during specified office hours to answer parent or student questions online. Suggestions to parents might help them propose contexts for the questions, for example, *something about counting* or *something about adding or subtracting* or *something about what the symbol looks like.*

Teachers could provide parents with some sample answers but should strongly discourage them from simply sharing those samples with their children. Assure parents that feedback will be provided and that getting the task "right" is not the focus, but that the responses provide the teacher with valuable feedback to know where to go next with students. Sample answers (i.e., sample questions with an answer of 5) to include for parents might be:

- What is 3 + 2?
- What number comes after 4?
- What numeral has 4 letters when it is written as a word?
- What is a number less than 10?
- What is half of 10?

Encourage students to buddy up with another student in the class (teachers might assign a buddy or not) to help students come up with a few ideas.

Students might submit their responses to the teacher through a quick video they make of themselves along with a snapshot of any pictures they might have created, perhaps the next day.

Feedback might focus on the number of mathematical ideas that students showed mastery of when creating their questions, or perhaps on the uniqueness of some of their questions. If students were asked for explanations of how they knew their answers were correct, teachers could provide feedback on the clarity of the explanations. Feedback might also involve extensions, such as asking students how they would change their questions if the answer were 6. Feedback might be provided individually by email to students, but could also be provided online for a particular group of students who would benefit from that.

> Make up an addition question where there is a 2, a 3, and a 4
> somewhere in the question or the answer.

A teacher could present the question in a form such as this:

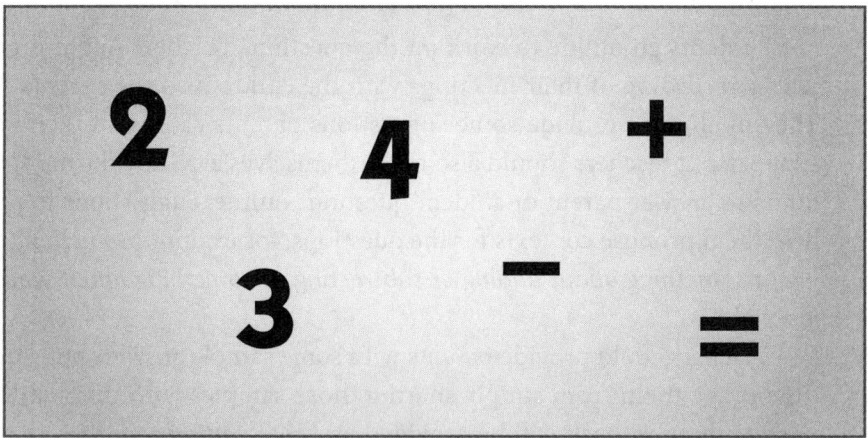

A teacher might provide quick online instruction in a class meeting, modeling for students an addition such as $5 + 1 = 6$ and pointing out that there is a 5, a 1, and a 6 in the question or the equation. Or a teacher could show how $5 + 5 = 10$ has a 5, a 1, and a 0 in it, and that $5 + 5 = 9 + 1$ has a 5, a 9, and a 1 in it. (Note that the examples deliberately do not include the values the students were asked to use but also deliberately model different forms of addition questions.) Finally, the teacher could bring the students back to their specific task: creating an addition question where there is a +, a 2, a 3, and a 4, and maybe other numbers and operation signs as well.

As in the previous example, the teacher should indicate to students there are lots of possibilities for responses. Reassure them that there is not a particular question that is the best one. (Note: This reassurance, which I have mentioned several times, is important since students often worry about what the teacher is hoping to get.) Students work on the task offline.

Suggestions to parents might help them scaffold the task should their children request help from their parents rather than their teacher. Suggested questions can be sent to parents by video. (An example video is available at tcpress.com/teaching mathonline, along with a script; the script is also available in Appendix A.) Probing questions might include:

- *What is your favorite addition? Write it down. If it does not have the numbers we need (2, 3, and 4), change one of your numbers to either 2 or 3 or 4. Now figure out the new answer. Do you have to change anything else to get in a 3?*
- *Could there be additions on both sides of the equals sign?*
- *Could there be addition on one side and subtraction on the other side?*

▶ **VIDEO:**   *Using Probing Questions:* 234 Question [Grades K–2]
Available at tcpress.com/teachingmathonline

Again, encourage students to buddy up with another student in the class to help them come up with a few ideas.

Students might submit their addition questions along with a discussion of how they started and what they did next and why.

A teacher could have a sharing meeting with groups of students online or simply provide individual feedback.

Feedback could focus on the strategies students used to make sense of the problem, an important standard of mathematical practice. For example, if a student wrote $2 + 3 + 4 = 9$, the teacher might ask whether it is possible for one of the 2, 3, or 4 to be on the other side of the equals sign.

> You add two numbers and end up with more than 7. What could you have added and how do you know? Could one of the numbers have been very small?

A teacher could present the question like this:

This particular question is probably quick enough and focused enough to complete with a small group of students in an online chat. A teacher might chat with a portion of the class at a time. The teacher might even consider grouping students by confidence in math and changing the number 7 to a higher number for some groups.

The teacher might provide quick modeling of how to add 2 + 3 on a number line. Point out, though, that the result of 2 + 3 is not more than 7 (or an alternate chosen number). Read the question to students and ask them to use different numbers (not 2 and 3) so that they end up with more than 7 (or that alternate value). Don't dwell on the issue of one of the numbers being small at first.

As in the previous examples, indicate to students there are lots of possibilities for sums greater than 7 (or the alternate value). Reassure them that there is not a particular question that is desired.

Have students talk about what they might do.

The teacher might ask questions like these:

- *Could the first number be 4? What would you add?*
- *Could it be 2? What would you add?*

The teacher might record student suggestions on an online white board for students to see. Once the list is complete, the teacher might ask if any of the additions involve a small number. Ask what students notice about the other number.

The purpose of this question is for students to notice that if one number is very small, the other one must be a lot bigger.

Feedback could focus on what students noticed about the relationship between a sum and the addends that make it up. How a teacher responds will, of course, be determined by what students offer. A student might, for example, indicate only that all the numbers were under 10 (since those were their only examples); the teacher could then ask if one of the numbers could have been more than 10. A student might, for example, say they know just because they know. The teacher could then ask what kind of picture that shows objects, not just numbers, they could draw to prove that they know.

> **_Choice 1:_** Choose a number to subtract from 20. Draw two different pictures that would help you figure out the result.
>
> **_Choice 2:_** Choose a number to subtract from 5. Draw two different pictures that would help you figure out the result.

The teacher could present the task like this:

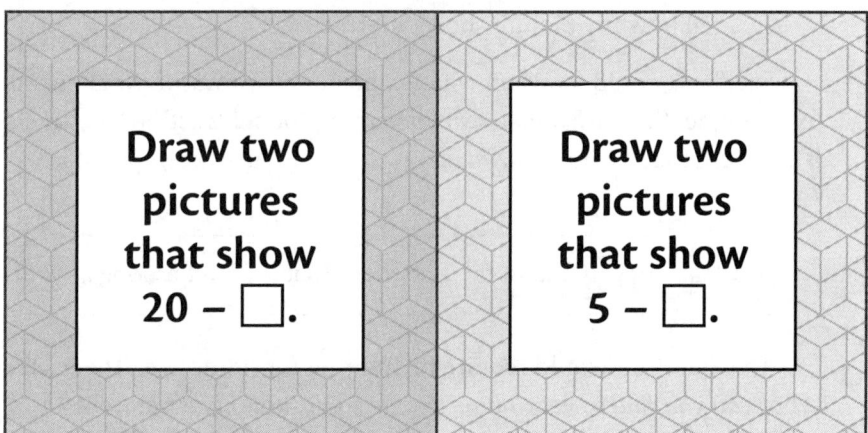

Emphasize that the pictures need to help someone figure out what the subtraction answer would be.

Parallel tasks can be useful in an online environment. Notice that both of these tasks require students to think about what subtraction means, but one of the subtractions involves smaller numbers. In a virtual environment, a teacher could choose to assign the appropriate task to the appropriate student and not make it a parallel situation, which is more critical when all of the children are together. Or a teacher might present both choices to students, reinforcing that students should choose just one of the tasks to do, not both. If both tasks are offered, the teacher might express the hope that some students pick the first question and some the second question to reinforce that one choice is not preferred over the other.

Make sure students understand that the pictures do not have to be fancy; students could even just use circles or squares, but the pictures can be more elaborate if students wish. Make sure, too, that students realize they cannot just write something like 5 – 2 or 20 – 2 since these responses do not really show that the student knows what subtraction means. Students must realize that somehow the picture needs to show what the minus sign really means. Students might also be allowed to send in a video of the subtraction happening instead of drawing a picture.

A teacher could suggest that if students wish, they could use objects around the house or could use ten frames or number lines.

For many sets of parallel tasks, *Good Questions* provides lists of common questions to ask students no matter which task they choose. These questions can be used virtually, as well, and the questions could be asked either online or offline. If offline, students could respond with a short video of themselves talking through what they did or, if that is difficult, they could send an audio text message.

Success criteria provided could be:

☐ I draw a picture to show subtraction.
☐ I tell how my picture shows subtraction.

Feedback could focus on what students show they know about subtraction. If, for example, they use a take-away meaning for subtraction, the teacher might ask if a picture that did not show take-away would have been possible.

What can you find in the classroom that is about as long as your arm?

This question would require re-framing for home use. The teacher might ask: *What can you find in your home that is about as long as your arm?*

This task does not require any complicated directions, so it could be presented without personal interaction with the students. Students could work on the task offline. To respond, students could either bring the item to a video chat with some or all students or submit photos that could be posted.

Success criteria might be provided, such as these:

☐ I find something as long as my arm.
☐ I tell how I checked that the length was right.

Feedback could focus on how students tested their object(s) to show that the lengths were correct, that is, their measurement process.

GRADES 3–5

> You multiply two numbers and the product is almost 400. What could the numbers have been? Explain your answer.

The teacher could present the task a little more attractively:

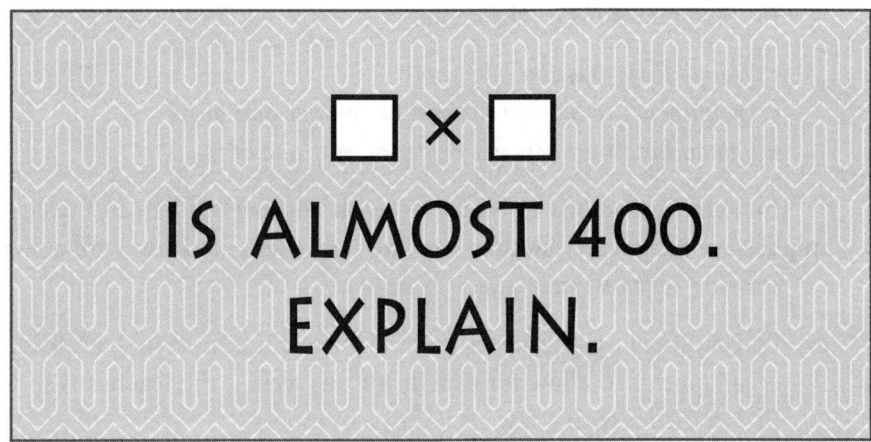

A teacher might send out this question digitally and then provide an opportunity for students to go online to ask clarifying questions, either in a group setting or through a digital message. Students are likely to ask:

- *How big do the numbers have to be?*
  And the teacher can indicate that this is truly the student's choice.

- *How close to 400 does the answer have to be?*
  And the teacher can, again, indicate that this is the student's choice.

- *Do we only need one answer?*
  And the teacher can indicate a preference, probably multiple answers in this environment.

- *Can the product be exactly 400?*
  And, again, the teacher can indicate a preference, allowing for this or not.

- *What do you mean by "explain"?*
  And the teacher can suggest, for example, a talk-aloud where students record all of their thinking as they work through the question.

Indicate to students that there are many possibilities for responding to this question. Reassure them that there is not a particular multiplication that is really desired, but that there are many good responses.

A teacher might provide parents with a recording of their meeting with the children so the parents can replay it, if necessary, for the children. Parents might also be offered other supports. A teacher could provide parents with some sample answers, but strongly discourage them from simply sharing them with their children. Instead, the teacher might share with the parents some probing questions and perhaps a few answers just to raise the parents' own comfort level, for example, 399×1, or 20×19, or 22×17, and so on.

Probing questions might include:

- *Would 10 × 10 work? Why not?*
- *Would you use bigger numbers or smaller numbers?*
- *Should both numbers get bigger or only one of them?*

Encourage students to buddy up with another student in the class (a teacher might have assigned a buddy or not) to help them come up with a few ideas.

Provide success criteria. For example:

- ☐ I multiply two numbers and the answer is almost 400. (A teacher could adapt this to allow for exactly 400, as well, if the teacher is so inclined.)
- ☐ I come up with a few sets of two numbers.
- ☐ I tell how I know my answers are right.
- ☐ I tell how I came up with my answers.

An additional criterion might be included:

- ☐ I tell what numbers not to bother trying and why.

The students go offline to work on the question. The teacher might collect individual or pair recordings of student thinking, or might propose a brief online meeting the next day, when lots of students could share their values.

If a follow-up chat occurs, rather than asking students to share answers for the questions they worked on, a similar question could be proposed where the product is 500. Students could respond on the spot since they had already completed the original similar problem.

Feedback might focus on the quality of student explanations of how they came up with answers. For example, did they start with a random question and adjust it, or did they start with a particular idea?

> Draw a picture that shows a division.

The teacher could present the task in a more eye-catching manner, as shown at the top of the next page:

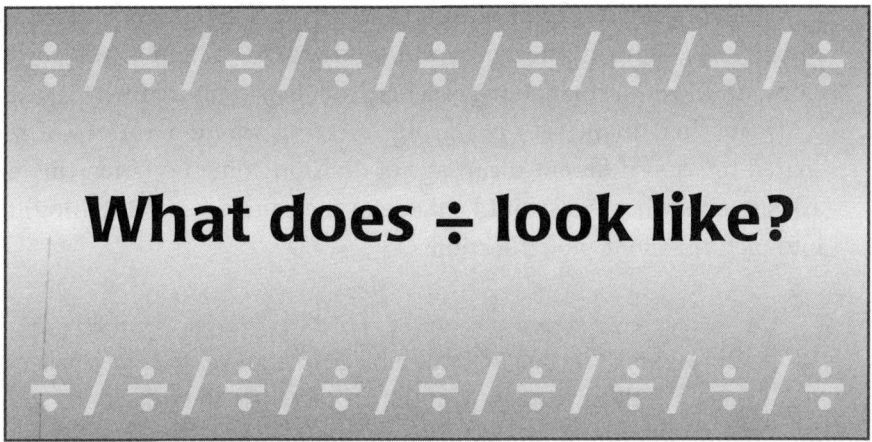

A teacher might provide quick instruction, sharing with students ideas about what a picture that shows addition might look like. (The point is to show students what it means to show an *operation* but not give away a response to the question about division.) Then students should be told that they are to draw a picture that somehow shows division.

Encourage students to think about different things division means and suggest that they create a variety of pictures showing those different meanings.

Let students know that the pictures need not be fancy, but that it should be clear why each picture is about division. Make sure, too, that students realize that the intent is not that they just write something like $10 \div 2$ or $20 \div 5$, since that does not really show what division means. Students need to understand that somehow the picture needs to demonstrate what the division sign really means. A teacher could also allow students to create a video of the division happening instead of drawing a picture. Students work on the task offline.

The teacher might provide parents with suggestions for questions they could ask their children if the children need support. These suggestions can be sent to parents by video. (An example video is available at tcpress.com/teachingmath online, along with a script; the script is also available in Appendix A.)

- *How would you figure out what $6 \div 2$ is?*
- *How could you show me that with these forks?*
- *Suppose you did a different division? What would be the same about what you show me?*

▶ **VIDEO:** *Using Probing Questions:* What Does Division Look Like? [Grades 3–5]
 Available at tcpress.com/teachingmathonline

Again, encourage students to buddy up with another student in the class to help them come up with a few ideas.

Students could take photos of their pictures along with explanations of why the pictures show division or could send videos of themselves performing divisions, again with explanations of where the division is and why it is division.

A teacher might set up a group meeting where a variety of responses are shared to see if different meanings of division come up. Then the teacher might ask students what they would change in their pictures (if anything) to turn them into pictures about multiplication.

> Draw a small rectangle. Draw a bigger rectangle that the smaller one is part of. Tell what fraction of the big rectangle the small one is.

The teacher might present the task more visually. For example:

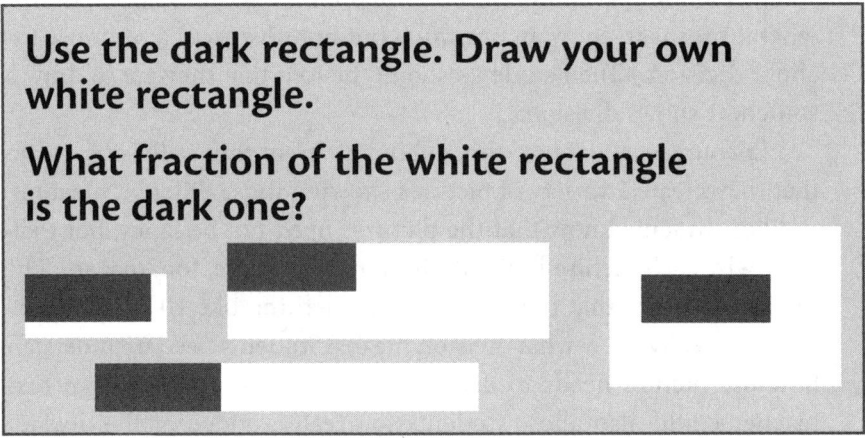

Use the dark rectangle. Draw your own white rectangle.

What fraction of the white rectangle is the dark one?

The purpose of this question is to bring out the mathematical notion that any object (or shape) can be any fraction of a different object or shape. Clarify that part of the white rectangle is under the dark one and that students do not have to use the white rectangles in the question but should create their own white rectangles. Given that students are at home and have time, a teacher might modify the task like this:

Draw a small rectangle. Make a bigger rectangle that the small rectangle is part of. Tell what fraction of the big rectangle the small one is.

Then do it again a few more times, but use different bigger rectangles to make different fractions.

For some (or all) students, a teacher might request that students create at least one picture where the fraction has a numerator of 2 or 3 rather than 1. The work is done offline.

Students could send their pictures in to the teacher digitally and the teacher could post a gallery accessible to all students in the class. At an online meeting with a group of students, the teacher might name one of the fractions and ask students to identify which picture she or he was describing.

Students might also be asked to explain how they created their own pictures and how they know that the fractions make sense. The latter discussion could happen individually for some students instead.

The focus of feedback might be on how students figured out the whole, as well as on whether their fraction makes sense.

> The areas of two shapes are almost the same, but the perimeters are very different. What might the shapes be?

Mathematically, the purpose of this question is to help students see that perimeter and area are independent measures of a shape, but, more specifically, that when two shapes have the same area, if one is much skinnier and/or has more "ins and outs," it has a greater perimeter.

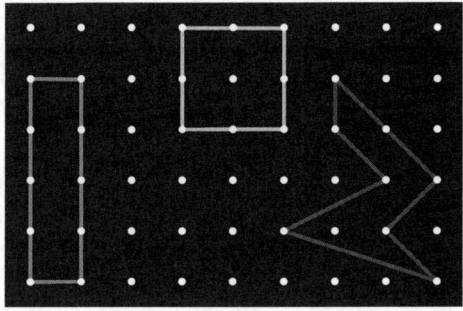

A teacher is likely to need a short instructional meeting with students in which the teacher shows a shape and clarifies what area and perimeter mean. The teacher might also suggest that students use dot paper or might provide a link to a geoboard app. (A template for dot paper is provided in Appendix B and at tcpress.com /teachingmathonline, and the Resources section lists sources for apps and online tools.)

▶ **MATH TOOLS:** Templates for Manipulatives
Available at tcpress.com/teachingmathonline

Remind students that they will need shapes with similar (but not necessarily identical) areas but very different perimeters. Clarify that the area they choose is up to them.

A teacher might provide success criteria such as these:

☐ I make at least 3 shapes with areas that are the same or almost the same.
☐ I make sure that at least one of the shapes has a really big perimeter.
☐ I make sure that at least one of the shapes has a really small perimeter.
☐ I explain how I know the areas are similar but the perimeters are not.

Give students a day to explore the task and have them submit pictures of their shapes. At a small-group chat the next day, choose a few submitted photos and have other students (not the ones whose photos were used) share how they know the areas are similar and how they know the perimeters are different. Having other students explain leads to more interaction.

A teacher could provide feedback by asking students to think about what it was about the shape with the bigger perimeter that made it bigger and then to record their response and submit it.

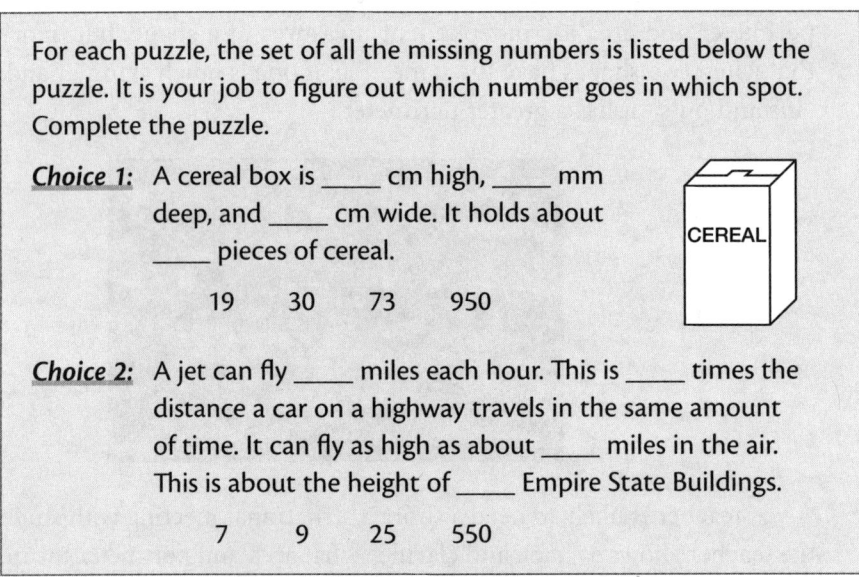

In a classroom, parallel tasks are valuable because there are different groups of students in the same room who need different questions to learn the ideas being explored. In the example shown here, both tasks involve recognizing when different sizes of numbers are used in real-world situations.

In a distance learning setting, a teacher could choose to give each student the one task the teacher believes is more appropriate for that student. However, the teacher might still offer both tasks in order to observe, if given a choice, which choice students make.

A teacher might need a short instructional video or an online meeting to clarify the task. Students are told that they need to put the four numbers listed at the bottom of the text for the question they choose into the four blanks, but not necessarily in the order given; that is, they have to figure out which number belongs in which blank and why.

Make sure students notice that some measurements are millimeters, some are centimeters, some are miles, and some are numbers not describing length units. Encourage students to try not to actually measure the cereal box if they select **Choice 1**, but to look at it. Students work on the task offline.

Students can be provided with these success criteria to use before they submit their responses:

☐ I put the 4 numbers where they belong.
☐ I tell which number I figure out first, then second, then third, and then last.
☐ I tell how I figure out which number belongs where.

A teacher could then provide a follow-up task where students create one of these puzzles for another student to do and submit it. Then each student could be provided with someone else's puzzle.

Feedback can focus on the strategies students used to place numbers and the inventiveness of their follow-up creation.

GRADES 6–8

> You multiply two integers. The result is about 50 less than one of
> them. What might the two integers be?

A teacher might choose a more visual representation of the problem. For example:

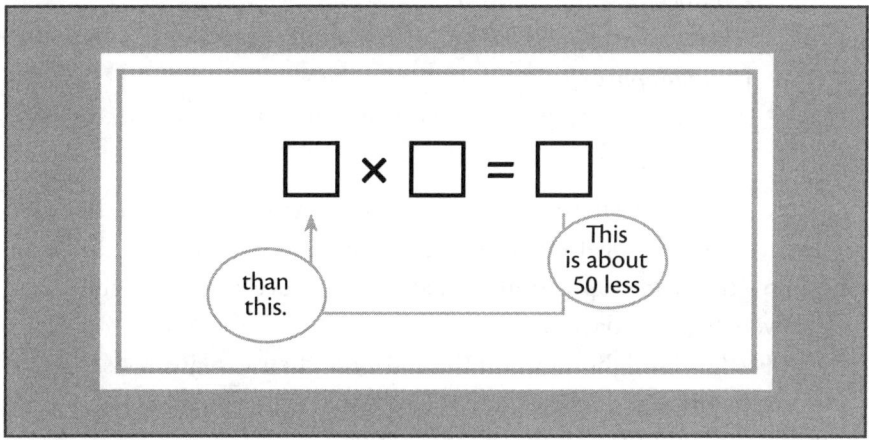

Before posing this question, a teacher might host a quick meeting where students are reminded of what an integer is and what being 50 less means.

For example, the teacher might show a number line with 0 marked, and tick marks with no other numbers, like this:

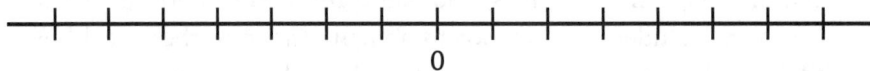

0

Then the teacher could point to spots at –2, +5, and –7 to see if students know what those integers are. Students could be asked what an integer 50 below –2 would be or what an integer 50 below +5 would be.

Ensure that students recall the sign rules for multiplying integers, for example, $(-3) \times (-2) = +6$ and $3 \times (-2) = -6$. Then post the problem students are to work on, reminding them that the result need not be *exactly* 50 less than either of the numbers being multiplied but just close to 50 less. Emphasize that it is the *product* that is less, and not the factor, and emphasize that it could be either factor.

Provide success criteria for students, such as these:

☐ I figure out different pairs of two integers where their product is about 50 less than one of them.
☐ I say what the product is and how I know.
☐ I explain how I came up with the integers I did.

Students work at the task offline. Students at this level may work more independently than younger students, but many are still likely to seek support from a parent. For that reason, parents might be provided with a recording of the meeting with the children so that they can replay it for students who need it. Parents might also be provided with suggestions for probing questions to ask to support students, such as these:

• *When you multiply 3 by 7, is the product more or less than the numbers you multiplied?*
• *When you multiply –3 by –7, is the product more or less than the numbers you multiplied?*
• *When you multiply 3 by –7, is the product more or less than the numbers you multiplied?*

Assign each student a buddy or allow them to choose buddies and submit their responses together. The focus of their submission should be an explanation of how they came up with the integers they did. Students might be asked to submit answers the next day.

Feedback might focus on the clarity of their explanations, particularly indications that they generalized and realized that unless they used 0, at least one integer had to be negative and the other positive.

> Create a parallelogram and a triangle so that the parallelogram area is half the triangle area.

If the problem is presented visually, it is important not to give away the answer, so rather than showing a possible picture, a teacher might choose to show a non-answer along with the actual problem. For example:

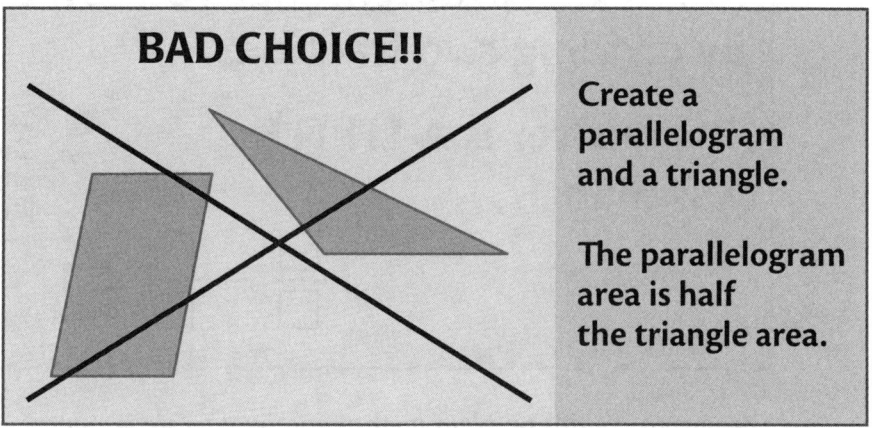

Set up an online meeting where students remind each other of what they know about the formulas for the areas of parallelograms and triangles. Make sure to provide those formulas with the problem as well.

Make sure students realize that it is the parallelogram area that has to be less this time, not the triangle area; this needs reinforcement since they are likely to think about the triangle area being less.

Reassure students that the actual values they use are up to them; the values could be small numbers or very large ones.

Provide some time for students to work on the question offline with a partner. Encourage them to come up with several pairs of shapes and even to generalize how this might be done by providing these success criteria:

- ☐ I choose areas for the parallelogram and triangle so that the parallelogram area is half as big as the triangle area.
- ☐ I tell what the base and height of each shape are and what the area is.
- ☐ I tell how I figure out the base and height values.
- ☐ I find lots of combinations.
- ☐ I describe a strategy for coming up with lots more possibilities.

Have pairs submit their work later the same day or the next day. Provide feedback with a focus on the strategies students used and, most particularly, their attempt to generalize. If you wish, you might follow up by asking students to share with you or with peers, either using pictures or talk, why the formulas make sense.

> You solve a problem that requires you to divide two fractions. The result is slightly less than 1. What might the problem be?

The teacher might present the problem more visually. For example:

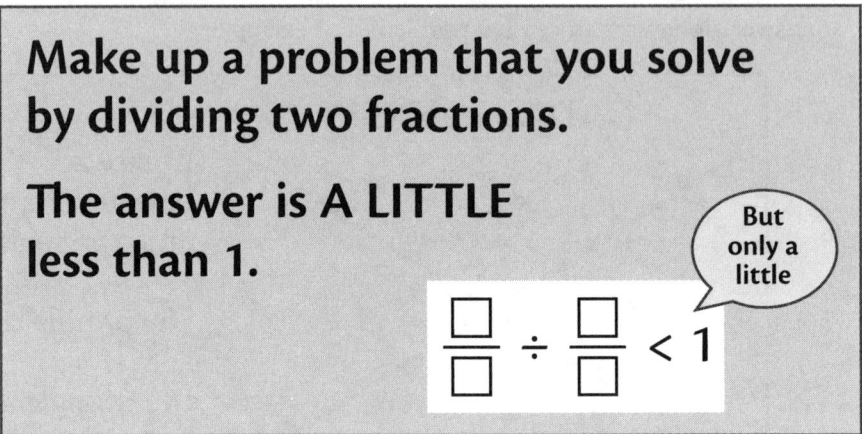

The purpose of this problem mathematically is to see if students realize what division of fractions means and recognize that since $\frac{a}{b} \div \frac{c}{d}$ tells how many times $\frac{c}{d}$ fits into $\frac{a}{b}$, the result is slightly less than 1 only if $\frac{a}{b}$ is slightly less than $\frac{c}{d}$.

One of the difficulties in using this problem online is that parents who help almost always think of division of fractions as inverting and multiplying rather than fitting one amount into another, and that might make the problem more challenging. As well, most parents would struggle with thinking of a real-life problem that leads to division of fractions. (Invariably, adults who are not math teachers seem to suggest problems that involve dividing by a whole number and not a fraction.)

This problem likely warrants a short online meeting with students to ensure they understand the task at hand. They should realize that the use of visual supports is welcomed. They should also understand that what is desired is not just an answer, but both a "story problem" and an explanation of how they figured out the answer. Students might be encouraged to generalize by rephrasing the question like this.

Part 1: You solve a problem that requires you to divide two fractions. What might the problem be?

Part 2: What must be true of the two fractions you divide to get a result slightly less than 1?

Students work on the problem offline. Encourage them to seek help during office hours or by email communication or text, and encourage them to work with a partner.

The teacher might send parents links to short videos of herself or himself both explaining when division of fractions is used (to find out how many of one amount fits into another or in determining a unit rate) and showing models students can use to calculate quotients of fractions.

Provided success criteria might include:

☐ I write a problem that would be solved by dividing two fractions.
☐ I make sure the answer to the division is almost 1.
☐ I show how I could use models to solve the problem.
☐ I describe what would be true of the fractions in any problem that works.

The teacher might post a number of submitted problems and ask students to choose a favorite and explain why they like that one. They might also be asked to create problems where the answers are slightly more than 1 or slightly less than 2.

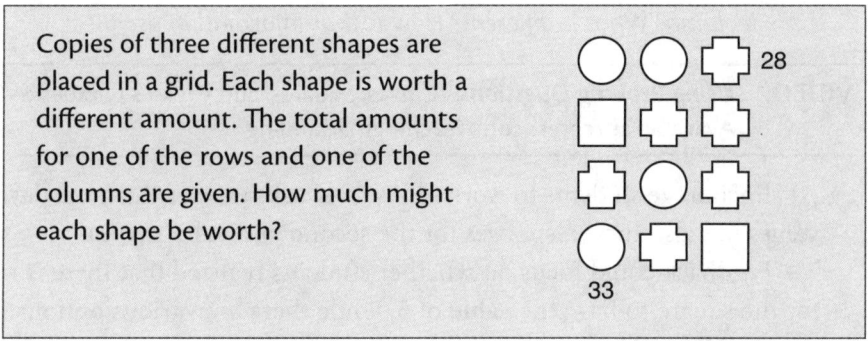

Copies of three different shapes are placed in a grid. Each shape is worth a different amount. The total amounts for one of the rows and one of the columns are given. How much might each shape be worth?

Although the task proposed here is algebra, it may not appear that way to either students or parents. It is not necessary to tell them that it's algebra, but a teacher might choose to signal that by requiring equations as part of the response.

A short instructional meeting can be offered to clarify what is being asked. Help students understand that once a value for a circle is selected, that value is maintained throughout the problem. The same is true for the value of a square or a cross.

Point out that information is provided only for the sums of the values in row 1 and column 1.

Provide success criteria. They might be, for example:

☐ I figure out values for the circle, cross, and square so that the total values given for the first row and first column are correct.

☐ I figure out if there are other possibilities or not. I explain why there are not if there are no other possibilities, and I explain how the different answers are related if there are other possibilities.

A teacher might provide a solution to parents to make the parents more comfortable should questions arise. Parents might also be shown that this is really algebra; the first row could be translated as $2x + y = 28$ and the first column as $2x + y + z = 33$ (where x represents the value of the circle, y the value of the cross, and z the value of the square).

Parents might also be provided with probing questions, which can be sent to parents by video. (An example video is available at tcpress.com/teachingmath online, along with a script; the script is also available in Appendix A.)

* *Suppose the circle were worth 1. Would you know the value of any other shapes?*
* *What's the same about the shapes in the first row and the shapes in the first column? What is different? How is that information useful?*

▶ **VIDEO:** *Using Probing Questions:* Circles, Squares, and Crosses [Grades 6–8]
Available at tcpress.com/teachingmathonline

Encourage students to work offline and submit later the same day not only the values but also their responses for the second success criterion.

Feedback could focus on whether students realized that there is no choice but for the square to have the value of 5, while there are various options for the other shapes. Note whether they point out that if the value of the circle increases, the value of the cross decreases, and whether they realize that the value of the cross has to be even if whole numbers are used. Note, too, whether they realize that if only whole numbers are used for answers, the maximum value for the cross is 28 (if the circle is 0) but the maximum for the circle is 14 (if the cross is 0).

> *Choice 1:* A number between 20 and 30 is 80% of another number. What could the second number be?
>
> *Choice 2:* A number between 20 and 30 is 150% of another number. What could the second number be?

The teacher might choose to present the problem more visually. For example:

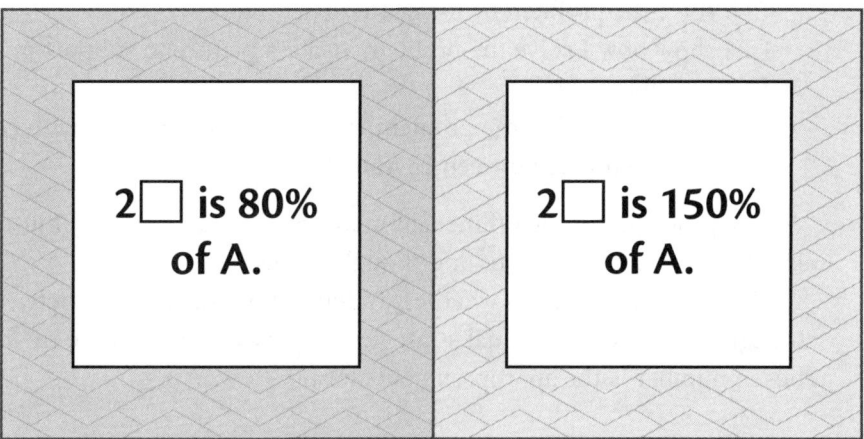

In a classroom, parallel tasks are valuable because there are different groups of students in the same room who need different questions to learn the ideas being explored. In a distance learning setting, a teacher could choose to give each student the one task the teacher believes is more appropriate for the student. However, the teacher might still offer both tasks to observe, if given a choice, which option students select.

A teacher might provide a short message in an online chat or by email ensuring that students realize that they are not to take 80% or 150% of something, but instead, after choosing the number between 20 and 30, they need to choose a different number that their number is 80% or 150% of.

Encourage students to use double number lines or 10 × 10 grids to support their thinking by providing hyperlinks to models that show them these tools. (The Resources section provides links to a variety of online tools. A video of an example lesson on use of double number lines is available at tcpress.com/teaching mathonline, along with a script; the script is also available in Appendix A. Templates for double number lines and 10 × 10 grids are provided in Appendix B and at tcpress.com/teachingmathonline.)

▶ **VIDEO & MATH TOOLS:**
Video: Learning About Math Tools: Using Double Number Lines [Grades 6 and 7]
Templates for Manipulatives
Available at tcpress.com/teachingmathonline

Students can be provided with these success criteria to satisfy before they submit their responses:

- ☐ I solve the problem by using a visual.
- ☐ I show how I solve the problem, using a picture to support my thinking if possible.
- ☐ I explain whether the problem would be easier to solve with certain number choices between 20 and 30 and why.

The problem is done offline. Submissions should provide not just answers but also the desired explanations and thought processes.

Feedback can focus on whether students chose the number between 20 and 30 strategically and on the visual strategies they used to determine the number their chosen number is the appropriate percent of.

SUMMARY

The problems selected and the strategies proposed here are examples of processes a teacher might think through in posing questions in an online environment in ways that will bring out desired mathematical concepts and processes. My hope is that adapting some of the suggested strategies to other tasks will make online learning easier and more productive for the teacher, richer for the students, and easier on parents.

CHAPTER 5

New Open Questions for the Online Environment

THIS CHAPTER provides a collection of some new open questions that particularly suit an environment where students have access to items in their home and when they have time to work on their own schedules without school bells and formal subject area changes that might normally curtail their investigations. These sorts of opportunities also go a long way in helping students see the value of math in their own lives. Teachers always strive to make math meaningful to students. When students' home environments become the content of their math learning, it has the potential to create a deep understanding that math is truly *everywhere*.

GRADES K–2

> What are there more than 5 of in your house? What are there fewer than 5 of?

A teacher might present the problem visually, for example, as below:

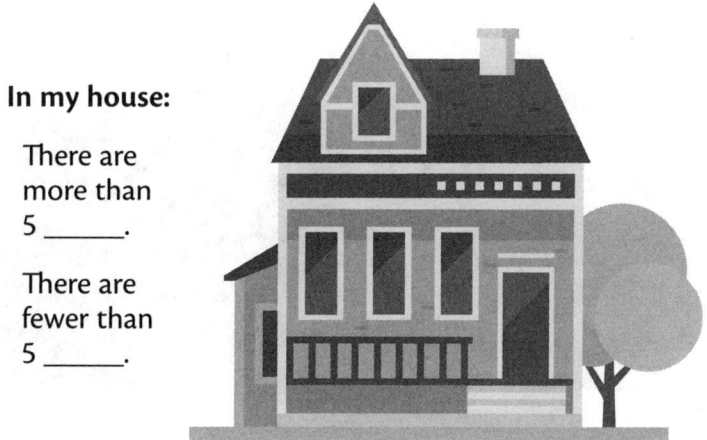

In my house:

There are more than 5 _____.

There are fewer than 5 _____.

Hold a brief meeting or send a message to parents to clarify what is expected (i.e., thinking of something that there are more than 5 of in their home and something there are fewer than 5 of). Let students know that they should think of lots of different possibilities, and not just one for each part. Allow students time to ask questions. For example, they might ask how many things they have to find for each list, and the teacher can clarify that they have freedom to decide. They might ask what to do if there are exactly 5 of something, and the teacher can suggest that they make another list if they wish, or that they can just not use those items. Students might ask how they should provide the information, and the teacher might suggest they get help taking photos of the items if they can.

The purpose of the question is twofold. The obvious purpose is to create meaningful opportunities to count. But, in addition, there is a measurement aspect. Since it is less likely that there are more than 5 copies of a really large thing in the house, and more likely that there are many smaller items, the teacher might follow up by asking which items were bigger—the ones there were more than 5 of or the ones where there were fewer than 5—and why they think that.

A teacher could take the opportunity to turn this task into a "puzzle," where a student is paired with another student. The first student names an item on one of the lists and the other student has to guess whether it was on the more-than-5 list or the fewer-than-5 list.

> Choose a number. Arrange toys to show that many toys.

Again, the teacher might present this task in a more interesting visual fashion by creating a pile of toys and then asking the question above.

The teacher might tell a story to introduce the problem in a little video sent to students. The story might suggest that a child who was visiting said that there is only one way to show 5 toys and you started to wonder if he was right. The teacher could then ask the students to think of their own number (that does not have to be 5) and explain whether there are lots of ways to arrange that number of toys or not. They could model arrangements.

The teacher might ask students not only to send photos or drawings of different arrangements, but also whether they think there would always be lots of possible arrangements or not, and why.

How high is a stack of 20 books?

This problem, too, can be presented visually:

Provide success criteria such as these:

☐ I make sure I have a pile of 20 books.
☐ I measure my pile and tell what I measured with and how many units I used.
☐ I explain why my answer might be different from someone else's answer.

Make sure that students and parents know that nonstandard measurement units (e.g., paper clips or Lego blocks or linking cubes) or standard units (such as inches or centimeters, depending on the age of the child) can be used. Make sure children and parents know, too, that the books need not be identical in size, and no particular type of book is required.

The last success criterion is at the center of the purpose for using this task. The objective is to help students see that when nonstandard units are used to measure something, or when there is variation in what is being measured, measurement values vary.

Make up a schedule using pictures of clocks to describe how you spend your day tomorrow. Include all the things you do.

While students are not in school, it becomes important to connect with them by allowing them to share their experiences. Asking them to create a schedule allows the teacher to talk personally with students, while providing opportunities for students to explore the mathematical topic of time.

Although there might be a temptation to provide an example schedule so that students know what is desired, it is highly likely that students will virtually copy the example and not really create their own schedule. For that reason, a teacher might have a short meeting to set up the task, ensuring that parents are provided with the same information about what is desired.

The teacher might suggest that the student look at the clock when first getting up and show that time on the schedule along with the words (or a picture showing): *I got up.* Then they could think about other things they do that take a little bit of time. For example, they do not have to write the time for going down the stairs, but they might write the time for eating breakfast, getting dressed, going out to play, and so on. It might be helpful to provide a template, such as this one:

My Day	
Time	**What I Did**

For students who are ready, the teacher might ask which of the things they listed took particular amounts of time, for example, about half an hour, about an hour, or about 15 minutes.

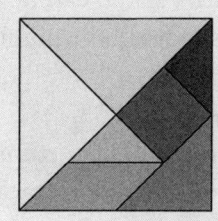 Use at least 3 of these 7 pieces to make one shape that represents your house or something in or around the house. Tell 4 things that are true about your shape.

Make sure that students have been given an outline for tangram shapes that they can cut out or, alternatively, provide a link or URL for a virtual tangram app. (A template for the tangram puzzle is available at tcpress.com/teachingmath online, and the Resources section lists sources for apps and online tools.) The fact that these pieces are called *tans* or *tangram pieces* might be shared, perhaps with some historical background about tangrams, but this is optional.

▶ MATH TOOLS: Templates for Manipulatives
Available at tcpress.com/teachingmathonline

Have a brief meeting with students or send a message to ensure that they know they can pick any of the pieces they wish to create their shape, but they may use each piece no more than once. Explain that the shape students form should be something that is made up of pieces connected at their edges and that they should be able to trace around it; it has an inside and an outside.

For example, these are shapes with pieces connected at their edges:

 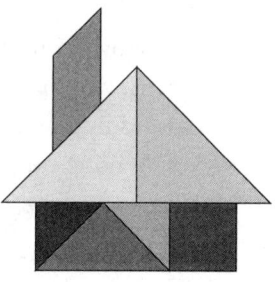

But this is not:

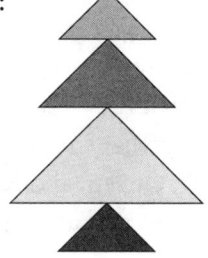

Clarify for students that the description of their shape can talk about many different features: for example, the number of sides (edges), the number of vertices (corners), the name of the shape, what other shape it is half of, the names of the shapes that make it up, another shape that it is like, and so on.

The value of the question is that it provides so much opportunity for discussion about a wide variety of attributes of shapes. For example, a teacher might ask:

- *Did anyone make a shape with more corners than sides?* (Through this question, students will discover that this is not possible.)
- *Did anyone make a shape that has sides that make a corner like a square?*

Make up a game with numbers or cards you can play with someone in your house.

The teacher might set up this task by having a short chat online with students telling them that the game could be about:

- Showing numbers different ways
- Moving on a 100-chart
- Adding or subtracting numbers on cards
- Making numbers using one or two cards

Explain that they get to make up the rules for the game, but there has to be a way for someone to win the game.

Again, rather than providing the child with examples, the teacher could answer some questions about the game that students might raise online. For example:

- *How many points do you need to win?* (Indicate that the child can decide, but it might be 5 points or 10 points or even 100 points.)
- *How many people can play?* (Indicate that the child can decide.)
- *How many cards should they use in the game?* (Indicate that it could be a deck of playing cards without the Jack, Queen, and King, or it could be as many number cards as they wish to make and as many copies of each number as they wish.)

Students might take some time to create a game, but they should be told they need to try it out with someone at home. Then they will describe their game in an online meeting.

Share games so that different students can play each other's games.

GRADES 3–5

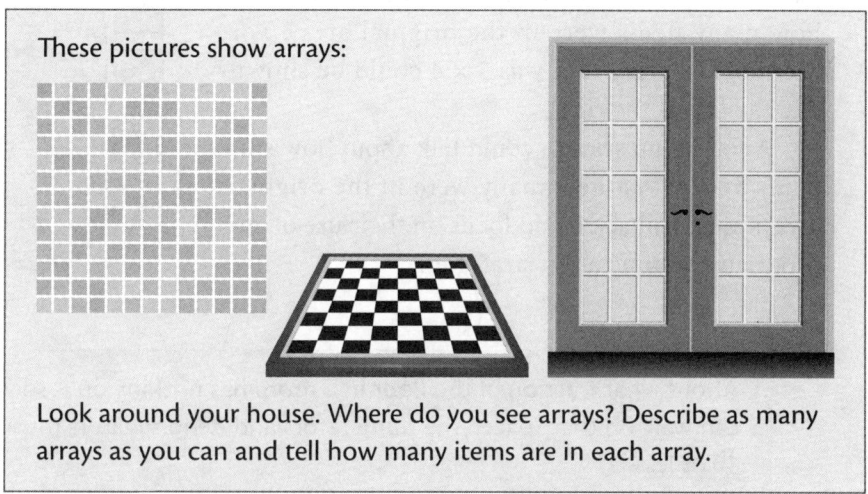

These pictures show arrays:

Look around your house. Where do you see arrays? Describe as many arrays as you can and tell how many items are in each array.

Arrays are an important multiplication tool for this grade band. Clarifying what arrays are and considering how to use them to make counting easier supports development in multiplication.

The teacher might introduce the idea of an array with a task where students decide what a number of items have in common and what distinguishes those items from others. For example, these pictures belong together:

These don't:

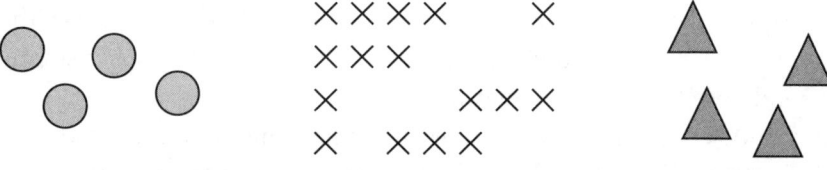

What is the difference between how the things are arranged in the first group and how things are arranged in the second group?

Once students are clear on what an array is, encourage them both to locate and take pictures of arrays in their home spaces or outdoors near their homes. They also need to describe how many items are in each array and to tell how they could figure it out without counting each and every item separately.

The teacher could then put together an online activity where the submitted photos are used, but part of the picture is hidden; students could guess how many items were in the original array. For example, an array that was 5 × 4 could be shown like the one at the right.

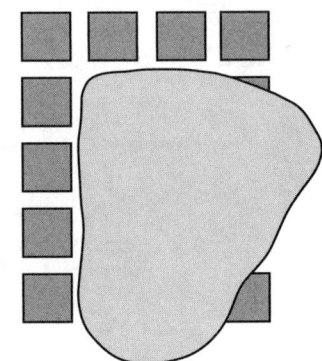

A follow-up session could talk about how students figured out how many were in the original arrays and feedback could focus on their use of appropriate multiplication strategies.

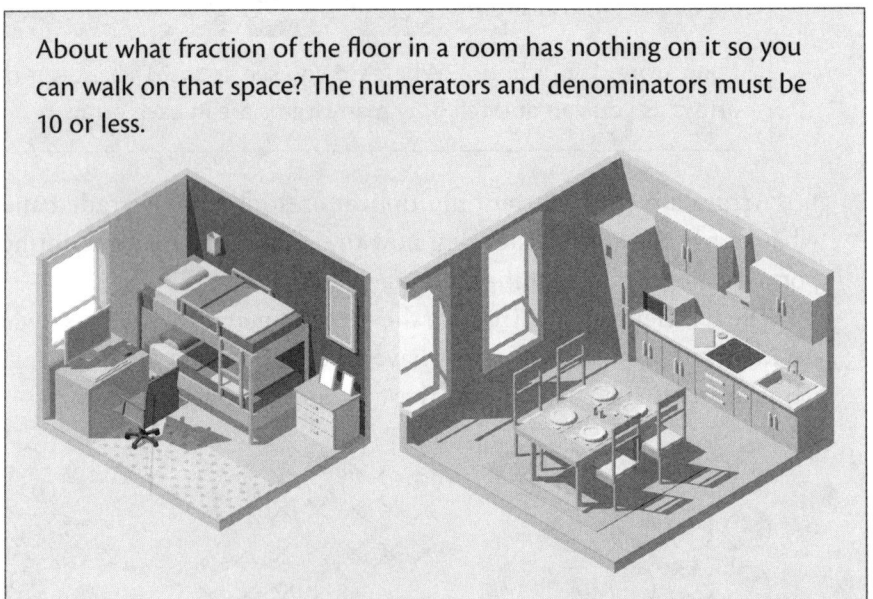

About what fraction of the floor in a room has nothing on it so you can walk on that space? The numerators and denominators must be 10 or less.

The teacher could present this task by beginning with an online discussion of rooms that are crowded with furniture and rooms that are not. Students can recall experiences they have had of these different kinds of spaces. Then the teacher could wonder out loud, *I wonder what fraction of the floor is empty in my kitchen,* and pose the provided question.

Ask students to share how they might measure their rooms to estimate the fractions. Then provide offline time for students to make the measurements and show how they determined the fraction estimates. Ensure that students realize that they need not use exact fractions, but instead can make fraction estimates that are easy to understand (i.e., those with numerators and denominators of 10 or less).

Provide success criteria. For example:

- ☐ I measure the amount of the floor that is covered.
- ☐ I measure the amount of uncovered floor.
- ☐ I tell how I measured.
- ☐ I create the fraction that compares the empty part to the whole area and estimate it with a fraction with numerators and denominators that are 10 or less.

The teacher might have students submit their work with drawings of the rooms (not necessarily to scale, but close) as well as clear descriptions of how they measured and how they estimated the fraction. Make sure students realize that to determine the uncovered floor area, they could have actually measured that area (probably by combining many smaller areas) or they could have measured the full floor area and subtracted the covered area. Make sure, too, that their fractions compare uncovered area to total floor area and not covered area to uncovered area or uncovered area to covered area. Follow up by asking for suggestions for fractions that would represent the amount of uncovered space in an average kitchen, an average dining room, or a very crowded living room.

How much space might you need to store 5 bags of rice?

At first glance, this might not seem like an open question, but it is. What is left open is the size of the bag, whether the unit of measurement is volume or area, and what units are to be used.

This particular task is easy to send home digitally without a pre-conference, but parents might wonder about the teacher's expectations for the issues that are left open. Parents will need support in understanding why the question is left open; they might be provided with some supporting probing questions, such as the ones that follow:

- *Would the needed space be different for different-sized bags?*
- *How could you figure out how big a bag of rice usually is?*
- *Do you think the space would be different if the bags were on top of each other or if they were not?*
- *What unit might you use to measure with? Why?*

Feedback could be offered individually to students by focusing on the appropriateness of the assumptions made, whether they realized they were making assumptions or not, their choice of units, and their recognition that the question could have been approached quite differently.

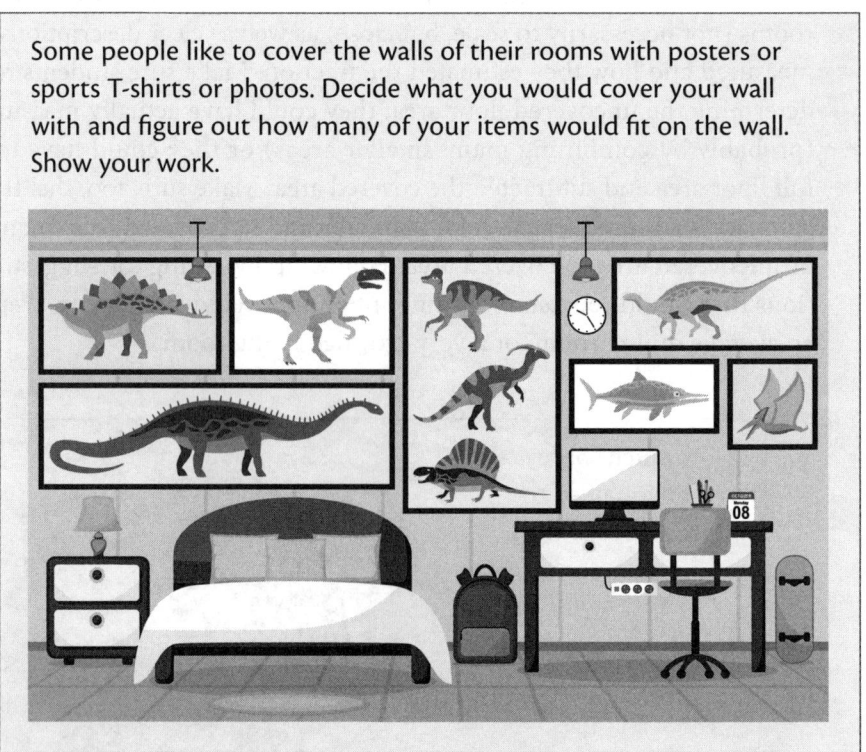

This particular task addresses both multiplication and measurement. It might pique student interest more to start with an online conversation with a small group of students asking if any of them or their siblings have lots of photos or posters or such things covering the wall in a room.

After assigning the task and gathering responses from students, a teacher might choose a result that is relatively small compared to others and one that is quite large. A follow-up conversation could provide the two extreme values to a group of students and students could hypothesize about what the items might have been and why they think that. After some hypotheses have been offered, the teacher might show the work that indicates what the items actually are.

> Keep track of how many times someone enters your kitchen each day. Estimate how long it would take until there are 100,000 entries to your kitchen.

This particular problem allows students to measure a length of time in whatever units they wish. They also need to come up with a manageable system for counting entries to the kitchen since they are unlikely to stand guard at the kitchen door all day.

The problem is certainly suitable while students are home. The teacher might encourage students to respond not only to the main question but also to these questions:

- *Did you assume the same number of entries every day of the week? Is that reasonable?*
- *Did you assume the same number of entries during different times of the year? Is that reasonable?*
- *Why do you think other students might end up with different lengths of time than you did?*

GRADES 6–8

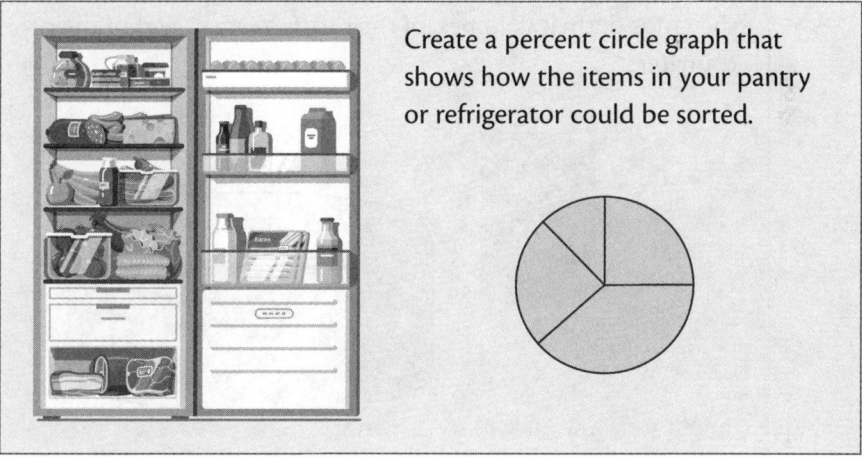

Create a percent circle graph that shows how the items in your pantry or refrigerator could be sorted.

The intent of the question is for students to consider how to use percentages to describe things. They get to choose the sorting criteria to make the task more or less complex. They might choose to consider type of food or location within the refrigerator. They might consider sizes of items, numbers of items, weights of items, and so forth, in making their choices.

Using a circle graph familiarizes students with a type of graph they are likely to see in the media regularly.

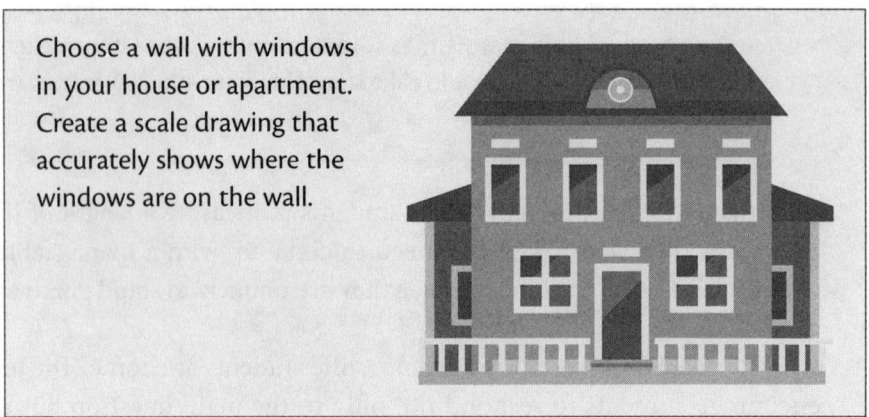

Choose a wall with windows in your house or apartment. Create a scale drawing that accurately shows where the windows are on the wall.

With students at home, having them create a scale drawing of their own space makes sense. But the mathematically important part of the task is that students describe their process and how they know that their drawing is accurate. That suggests making sufficient actual measurements to ensure accuracy and including measurements as well as appropriately scaled windows on their drawings.

The focus of the discussion should be on how the student knows the drawing is a correct scale drawing.

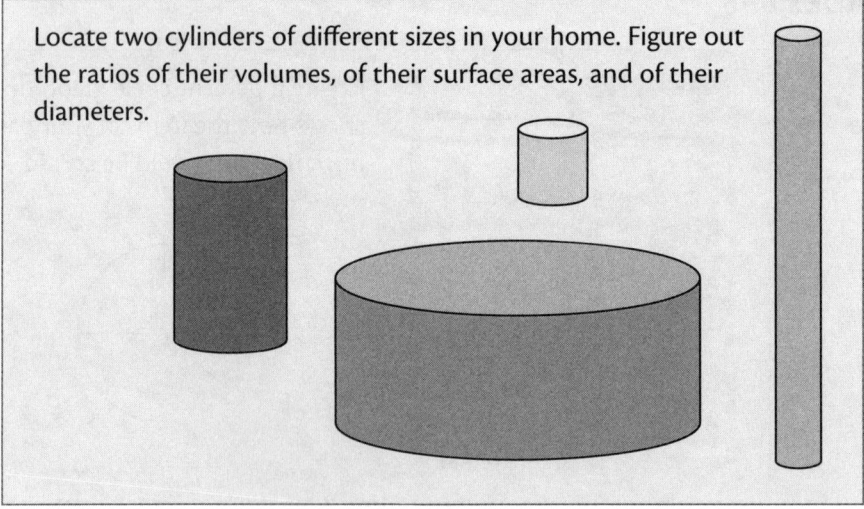

Locate two cylinders of different sizes in your home. Figure out the ratios of their volumes, of their surface areas, and of their diameters.

Students are asked to consider a number of different ratios as they apply the formulas for volume and surface areas of the cylinders. They are apt to discover that the ratios of diameters, surface areas, and volumes of two cylinders differ. The task will also help students notice where cylinders regularly appear in their everyday world; they might be soup cans or lamps or pipes or swimming pools or a sibling's toys.

This question is clear enough to present to students to complete without any introduction, but could be followed up with questions like these:

- *What might the cylinders' dimensions be if the ratio of their volumes is close to 5:1? What would their surface area ratio be then?*
- *What might the cylinders' dimensions be if the ratio of their surface areas is close to 5:1? What would their volume ratio be then?*

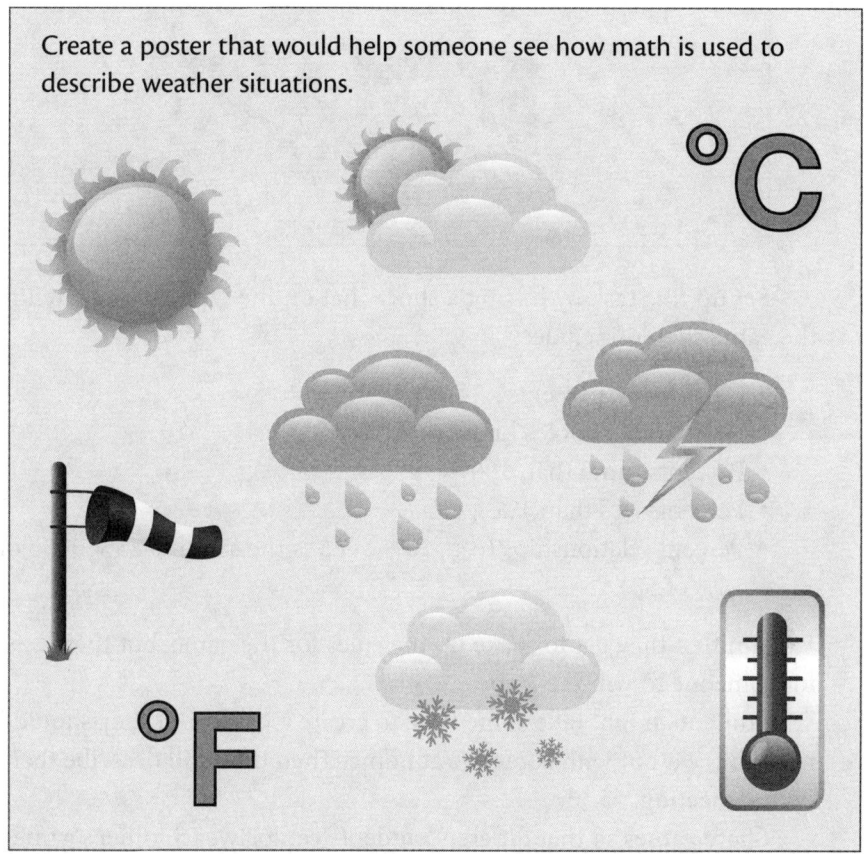

Create a poster that would help someone see how math is used to describe weather situations.

The question posed here might lead to an extremely interesting investigation for students who are interested in weather reports. They would have the time in a virtual environment to do lots of online investigation. They could be encouraged, through success criteria, to include situations involving:

- Percents
- Negative integers
- Probability
- Measurement units
- Data distributions

Create a game involving percents. Test it with someone else. Describe the rules and explain its value as a math game.

Set up this task by having a short chat online with students telling them that the game should include:

- Calculating percents
- Calculating wholes knowing percents
- Percents more than 100%
- Percents less than 1%
- Percent relationships (e.g., 30% of 25 is the same as 25% of 30 or 15% of 50).

Explain that they get to make up the rules for the game, but there has to be a way for someone to win the game.

Students might take some time to create a game, but they should be told they need to try it out with someone at home. Then they will describe their game in an online meeting.

Share games so that different students can play each other's games.

SUMMARY

These are just a few examples of appropriate distance learning activities for math. There are many other opportunities to use open questions when students have access to home resources and more unstructured time than usual.

❖ CHAPTER 6 ❖

Conclusion

AS TEACHERS engage with the challenges of distance or blended online teaching, it can be helpful to keep in mind some key principles for teaching mathematics. While teaching in an online situation can be challenging, teachers may find that there can be some enrichment arising from the personalizing of mathematical questions for different students and from relating work with math to the home environment that will pay dividends in terms of students truly grasping that math is all around them.

In closing, I wish to reinforce a few points I have already mentioned that teachers should keep in mind regarding the big differences in using open questions or parallel tasks online:

- Activities need to be visually appealing.
- Teacher presence in video meetings needs to be especially positive and energetic.
- Posing of questions and activities to be completed needs to be reasonably concise.
- Students need clear indications of what is expected, including success criteria, and, at least sometimes, of whether there are many correct answers or not.
- Students need clear indications, at the outset, of whether work is to be turned in or not, and, if so, when and how.
- Students may need help in finding a "buddy" to work with.
- Quick virtual meetings to clarify what is being asked often must be provided, though not always.
- Availability of materials for students to work with must be considered carefully.
- Parents need to be provided with supporting questions, particularly for students in Kindergarten through Grade 5, and with schedules or overviews of what will be expected of students, when, and how much time will be required.
- Simple platforms are needed for teachers delivering questions, but also for students responding to activities.
- Techniques for monitoring engagement are essential.
- Feedback must include not only text but also some interactive meetings.

Resources

REFERENCES

Council of the Great City Schools. (2020). *Addressing unfinished learning after COVID-19 school closures.* https://www.cgcs.org/cms/lib/DC00001581/Centricity/Domain/313/CGCS_Unfinished%20Learning.pdf

Darby, F. (2019). *How to be a better online teacher: Advice guide.* The Chronicle of Higher Education. https://www.chronicle.com/article/how-to-be-a-better-online-teacher

Huinker, D., Yeh, C., Rigelman, N., & Marshall, A. M. (2020). *Catalyzing change: Overview of the 4 key recommendations for early childhood & elementary mathematics.* https://www.nctm.org/uploadedFiles/Conferences_and_Professional_Development/Webinars_and_Webcasts/Webcasts/Catalyzing-Change-Overview-Webinar-Final.pdf

Puentedura, R. R. (2012). *The SAMR model: Six exemplars.* http://www.hippasus.com/rrpweblog/archives/2012/08/14/SAMR_SixExemplars.pdf

Small, M. (2017). *Teaching mathematical thinking: Tasks and questions to strengthen practices and processes.* Teachers College Press.

Small, M. (2019). *Math that matters: Targeted assessment and feedback for Grades 3–8.* Teachers College Press.

Small, M. (2020). *Good questions: Great ways to differentiate mathematics instruction in the standards-based classroom* (4th ed.). Teachers College Press.

Vygotsky, L. S. (1978). *Mind in society: The development of higher psychological processes.* Harvard University Press.

ONLINE RESOURCES

Didax: https://www.didax.com/math/virtual-manipulatives.html

Flipgrid: https://info.flipgrid.com

GeoGebra: https://www.geogebra.org/m/NPDu3rCm

Mathies (sponsored by the Ontario Ministry of Education): https://www.mathies.ca/#gsc.tab=0

Mathigon's Polypad: https://mathigon.org/polypad

The Math Learning Center: https://www.mathlearningcenter.org/new/blog/topic/math-manipulatives

Math Playground: https://www.mathplayground.com/math_manipulatives.html

National Library of Virtual Manipulatives (Utah State University): http://nlvm.usu.edu/en/nav/vlibrary.html

The Techie Teacher: https://docs.google.com/presentation/d
/1jadIg9nk64U9gWtj4QEbd-AEzVtCXQeH-y44LPXlF3M/present?slide
=id.g27b693dca5_0_261

Toy Theatre: https://toytheater.com/category/teacher-tools/virtual-manipulatives

VIDEOS

Eight videos that have been prepared as models of videos that can be sent to parents may be accessed from the Teachers College Press webpage for this book (tcpress.com /teachingmathonline). Five of these (**Learning About Math Tools**) focus on applying important math tools in the online environment, and three (**Using Probing Questions**) illustrate the use of probing questions in stimulating rich math conversation. Scripts for the videos appear in Appendix A and are available online. Instructions for accessing the videos and scripts may be found in Appendix A.

Learning About Math Tools

Using a 100-Chart [Grades 1 and 2]
Using Number Lines [Grades 2 and 3]
Using Pattern Blocks [Grades 3 and 4]
Using Base Ten Blocks [Grades 4 and 5]
Using Double Number Lines [Grades 6 and 7]

Using Probing Questions

234 Question [Grades K–2]
What Does Division Look Like? [Grades 3–5]
Circles, Squares, and Crosses [Grades 6–8]

 VIDEOS: Available at tcpress.com/teachingmathonline

MATH TOOLS

Appendix B provides templates for a number of manipulatives discussed in the text. Those templates, plus others, are available as a downloadable and printable PDF file at tcpress.com/teachingmathonline.

 MATH TOOLS: Available at tcpress.com/teachingmathonline

❖ APPENDIX A ❖

Scripts for Example Videos

THE VIDEO SCRIPTS included in this appendix correspond to videos available for viewing on the Teachers College Press webpage for *Teaching Math Online*. These videos have been prepared as examples of videos that can be sent to parents, and the scripts are provided as models for teachers who would like to use them as bases to create their own videos or online lessons. The first five scripts (**Learning About Math Tools**) demonstrate use of common manipulatives for lessons on a variety of topics at grade levels K through 8. An additional three scripts (**Using Probing Questions**) demonstrate ways parents can support students in their learning with examples of probing questioning on specific math tasks.

The videos are available for viewing on the Teachers College Press website and are not downloadable. To view, go to tcpress.com/teachingmathonline, select the video you wish to view, and enter the following password at the prompt:

Password: !qANm<:

Password Required
This content requires a password to view: [   ] Send

Scripts for the videos are provided in this appendix, as listed below, and also appear at tcpress.com/teachingmathonline.

Learning About Math Tools	Using Probing Questions
Using a 100-Chart [Grades 1 and 2]	234 Question [Grades K–2]
Using Number Lines [Grades 2 and 3]	What Does Division Look Like? [Grades 3–5]
Using Pattern Blocks [Grades 3 and 4]	Circles, Squares, and Crosses [Grades 6–8]
Using Base Ten Blocks [Grades 4 and 5]	
Using Double Number Lines [Grades 6 and 7]	

▶ **VIDEOS:**　Available at tcpress.com/teachingmathonline

LEARNING ABOUT MATH TOOLS
Using a 100-Chart [Grades 1 and 2]

▶ **VIDEO:**   *Learning About Math Tools:* Using a 100-Chart [Grades 1 and 2]
                Available at tcpress.com/teachingmathonline

I am going to show you the **100-chart**. It's such a valuable tool for your children for
  comparing numbers.

One of the things they learn is that if a number is higher on the chart, it is worth
  less. So let's look at 23 and 45. 45 is lower, so that means that 45 is greater. And
  that also means 23 is less.

1	2	3	4	5	6	7	8	9	10
11	12	13	14	15	16	17	18	19	20
21	22	(23)	24	25	26	27	28	(29)	30
31	32	33	[34]	35	36	37	38	39	40
41	42	43	44	(45)	46	47	48	49	50
[51]	52	53	54	55	56	57	58	59	60
61	62	63	64	65	66	67	68	69	70
71	72	73	74	75	76	77	78	79	80
81	82	83	84	85	86	87	88	89	90
91	92	93	94	95	96	97	98	99	100

We can also compare numbers in the same line. Let's look at 23 and 29. Because
  23 is to the left of 29, it is less.

The idea is that down means more and right means more, but down matters the
  most. Let's look at 51 and 34. Even though 51 is to the left, it is down, and down
  means more.

I think this really helps kids see that the tens digit is what really matters the most in
  deciding which two-digit number is greater, since the tens digit tells you what
  line it's in. If it's a greater tens digit, it is lower on the chart.

LEARNING ABOUT MATH TOOLS
Using Number Lines [Grades 2 and 3]

▶ **VIDEO:** *Learning About Math Tools:* Using Number Lines [Grades 2 and 3]
Available at tcpress.com/teachingmathonline

Let's look at **number lines**. They are particularly useful tools for kids to do operations.

I am going to show you an example of a subtraction. This one is 512 − 415.

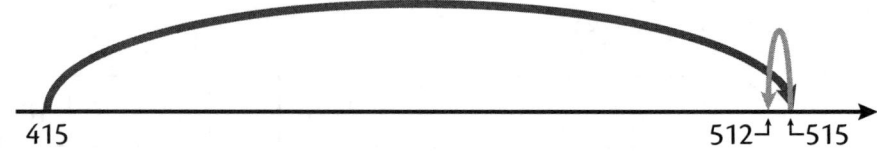

What you are noticing is that I am interested in: *How long is the jump to get from 415 to 512?*

That's exactly what 512 − 415 means.

Notice it's easy to add 100 to 415, but that gets you to 515, not 512.

So you still have to go back 3 to get to 512.

What I think is you've moved forward 100 and back 3. So you've moved 97, and that's the answer.

LEARNING ABOUT MATH TOOLS
Using Pattern Blocks [Grades 3 and 4]

▶ **VIDEO:** *Learning About Math Tools:* Using Pattern Blocks [Grades 3 and 4]
Available at tcpress.com/teachingmathonline

Pattern blocks are tools that we use with children to help them work better in geometry, but also in fractions.

Have a look at the shapes I am showing you here. There is a yellow hexagon, a red trapezoid, a blue parallelogram, and green triangles.

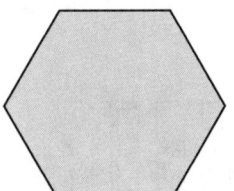  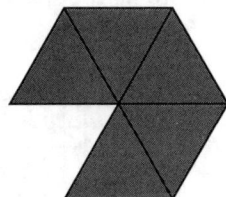

It turns out that two of the red fit on the yellow, so each red is worth $\frac{1}{2}$ if the yellow is worth 1. Three of the blue fit on the yellow, so each blue is worth $\frac{1}{3}$ if the yellow is worth 1. Six green fit on the yellow, so each green is $\frac{1}{6}$ of the yellow, so the green is called $\frac{1}{6}$ if the yellow is worth 1.

Now because of this fractional idea, we can actually show fraction calculations with the pattern blocks.

Let's show $\frac{1}{2} + \frac{1}{3}$.

The $\frac{1}{2}$ is the red, the $\frac{1}{3}$ is the blue. So $\frac{1}{2} + \frac{1}{3}$ means stick those together to see how much of the yellow you have covered.

You can see it's almost a whole yellow, but you're missing a little piece. You are actually missing $\frac{1}{6}$. That makes this $\frac{5}{6}$.

But you could also show this is $\frac{5}{6}$ by using five $\frac{1}{6}$ and seeing it's exactly the same size.

So $\frac{1}{2} + \frac{1}{3} = \frac{5}{6}$.

LEARNING ABOUT MATH TOOLS
Using Base Ten Blocks [Grades 4 and 5]

> ▶ **VIDEO:**   *Learning About Math Tools:* Using Base Ten Blocks [Grades 4 and 5]
> Available at tcpress.com/teachingmathonline

You might use **base ten blocks** to multiply. We can use them for lots of things, but let's look at them for multiplying 24 by 14.

Since 24 × 14 is the area of a rectangle that is 24 units by 14 units, you could build a rectangle with blocks, and then you could count the value of the blocks you use to fill the space efficiently.

24 units long (20 + 4)

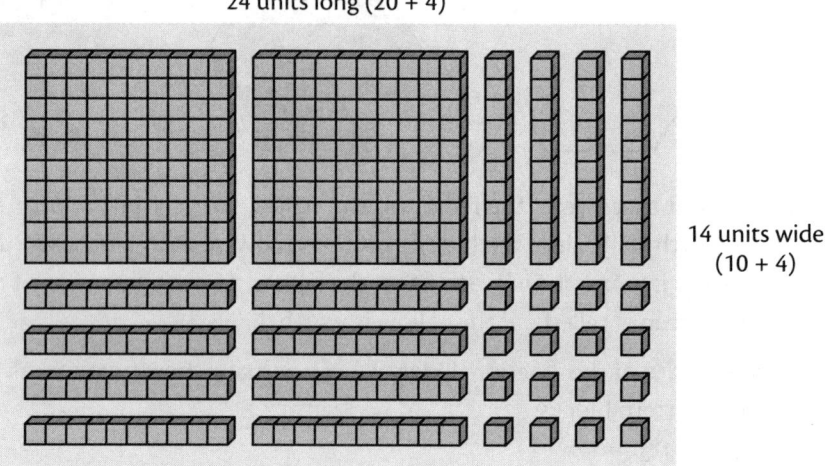

14 units wide
(10 + 4)

You can see here I am going along the top width of 24 and a length of 14, or you could call 14 the width and 24 the length. And I am filling the space efficiently, and that's using as many big blocks as I can.

So the area is how much stuff I use in here.

You can see I that I use 2 hundreds. You can see I use 1, 2, 3, 4, 5, 6, 7, 8, 9, 10, 11, 12 blocks that are ten, so that's another 100. And some ones blocks.

I am reinforcing that the product is 2 hundreds + 12 tens + 16 ones, which does give you the right answer of 336.

But it also reinforces estimation, since you can see when you look at the area that the big stuff is over here and maybe some of this over here. So an estimate of 300 for 24 × 14 makes sense. There's 100, 200, and about another 100.

The model is great, since it reinforces estimation as well as helps you with calculation.

LEARNING ABOUT MATH TOOLS
Using Double Number Lines [Grades 6 and 7]

▶ **VIDEO:** *Learning About Math Tools:* Using Double Number Lines [Grades 6 and 7]
Available at tcpress.com/teachingmathonline

Let's look at another model that I think is useful. This is something you can do on paper. You don't need any plastic materials or even cutouts.

And it is called the **double number line** and is really useful to show what I'm going to call ratio relationships, and that includes percents.

Imagine you have just solved a problem, and the problem is: *There was a sale and it was 40% off and you paid $36. And you want to know what the original price was.* So the sale price is $36, and you want to know the original price and you know it was 40% off.

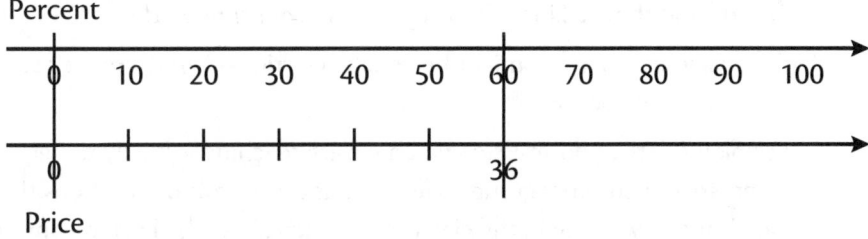

So what I am going to use is something called a double number line. And what you can see is that the top line shows percents, and it goes from 0 to 100. And the bottom line shows prices.

And you know that you had a 40% off sale, so you paid 60% and you paid $36.

You know that if the price had been $0 to start with, that would have been 0%. So what you are trying to do is to figure out what was the original 100% price.

You know the 60% price, and you want to know the 100% price.

Notice that you can see that there were 1, 2, 3, 4, 5, 6 steps to get to 36. That means every step is worth $6.

That means every 10% was worth $6, so you can use the 6 times table: 6, 12, 18, 24, ... And so 6 tens, which is 10 steps, gets you to 60. So the original price must have been $60.

It's a great way to show percents.

I know it is quick, you think, to calculate, but this actually makes sense—and why the answer is the $60 it really is—and making sense is always a good thing.

USING PROBING QUESTIONS
234 Question [Grades K–2]

▶ **VIDEO:**   *Using Probing Questions:* 234 Question [Grades K–2]
            Available at tcpress.com/teachingmathonline

> Make up an addition question where there is a 2, a 3, and a 4
> somewhere in the question or the answer.

We have a problem that we've assigned to kids where we have asked for an addition equation that has a 2 somewhere, a 3 somewhere, and a 4 somewhere. They can be anywhere. They could be in the what-you-add part, they could be in the answer part, but somewhere a 2, somewhere a 3, and somewhere a 4.

That's kind of a strange question for some kids. They don't know exactly what you mean, and they're kind of worried: *What am I supposed to do?*

And actually their parents might be equally worried—*What are they supposed to do?*—and not be sure.

So I think it's part of our mission when we are teaching virtually to help the parents who are our support group to learn to ask the kinds of questions that teachers are hopefully asking in the classroom, which allow the kids to continue to do the thinking, and don't do the thinking for them, and send them in the right direction.

I think it's good to send little videos where we offer some of these probing questions. So, for example, if I were doing the question I was just talking about, here are some of the questions I might put in a little video.

Something like: *You could ask questions like these to move your kid along.*

- A child might be asked: *Write down your very favorite addition. What is it?*
  - If a kid happens to write something that has a 2 or a 3 or a 4 in it, that would be great.

    Then you could say: *I see a 2, but I don't see a 3 or a 4. Could we change one of those numbers to a 3 or a 4? And what else would you have to change?*

    And you could kind of go in that direction.
  - Or it might be they wrote 5 + 5 = 10, which has no 2, no 3, and no 4.

    So you could say: *My favorite is 2 + 5 = 7. Ooh—I got the 2 in, but I don't have the 3 or 4. Can you help me get the 3 or 4?*

  I'm sure that's a direction you could go. Start with something they do.

- It could be something like: *I wonder if you could have addition on both sides of the equals sign, like something + something = something + something, but there's somewhere a 2, somewhere a 3, and somewhere a 4.*

  Yeah! Let's do that!

  And don't even say: *Can we?* Just say: *Let's do it.*

- Or you could say: *I wonder if you could put an addition on one side and a subtraction on the other side. I wonder if that would work.*

And use those kinds of questions to get kids to do thinking, sending them in the right direction—leading to positive mathematical performance and getting everyone through the moment in a bearable way.

USING PROBING QUESTIONS
What Does Division Look Like? [Grades 3–5]

▶ **VIDEO:**   *Using Probing Questions:* What Does Division Look Like? [Grades 3–5]
                 Available at tcpress.com/teachingmathonline

> Draw a picture that shows a division.

The problem for students is to tell me or to show me what division looks like, and that might be a hard question for them to figure out. *What does she want or what does she mean?* And even parents might wonder: *What do I want or what do I mean?*

It's good to equip parents with some of what I am going to call **probing questions** that they could ask their children to get them started without telling them the answer.

So let's think of something simple like 6 divided by 2. So they could go to their drawer and pull out 6 forks. *There are 6 forks here.* And say to the kids: *If I said 6 divided by 2, what do you think that means?*

And kids might say different things. One kid might say: *You have to share them, so 3 for you and 3 for me.* Or another kid might say: *It's 2 forks and 2 forks and 2 forks, so the answer is 3.*

Whatever they do is great, but what the parent is doing is getting the kid to think about what division means.

Then they might follow up by: *Suppose it wasn't 6 divided by 2. Suppose it was a different division. What would be the same about your story or what you did with the forks? Or what would be different?*

And that should help parents figure out how to help their kids without giving it all away.

USING PROBING QUESTIONS
Circles, Squares, and Crosses [Grades 6–8]

▶ **VIDEO:** *Using Probing Questions:* Circles, Squares, and Crosses [Grades 6–8]
Available at tcpress.com/teachingmathonline

Copies of three different shapes are placed in a grid. Each shape is worth a different amount. The total amounts for one of the rows and one of the columns are given. How much might each shape be worth?

Students at this age may or may not be involved in asking for help from their parents. I know 13-year-olds who ask their parents for help and others who don't.

Let's just say we have a student who gets a problem online from you like the one I proposed here with the circles and the squares and the crosses. And they don't know what to do, so they go to their Mom or they go to their Dad and they ask them some questions.

Parents are not used to these kinds of problems and the kinds of things they would do, so one of the things parents might do is go and try to figure it out themselves and then just give kids the answer to the question.

I think we need to share with them, just like we would parents of children of any age group, what are the kinds of questions teachers are asking kids to get them to think for themselves but to kind of understand better what the question is asking so they could help.

So here is a list—and I might do this in a video like I'm sharing with you now—of some quick questions you could ask your child.

- Here's one: *What if the circles were worth 1? So let's just try. The circles are going to be worth 1. Put those values in. Would you know anything else? No.*

  Then hopefully they might notice that in the first line, where there are two circles and a cross, they kind of do know what the cross would be worth. Then go from there.

- Or you could ask this: *What's the same about the shapes in the first row and the shapes in the first column? And what's different? So how would that be useful because you do know the first row totals 28 and you do know the first column totals 33? How might that be useful?*

Those are two just quick little examples of the kinds of questions you could share with parents that they could ask their kids that would keep children thinking but still move them in a positive but healthy direction to get to the answer.

Templates for Manipulatives

These 11 templates, along with others, are also available as a downloadable and printable PDF file at tcpress.com/teachingmathonline. Among the additional materials available online are the following:

- Addition Table
- Multiplication Table
- Tangram Puzzle

▶ **MATH TOOLS:** Available at tcpress.com/teachingmathonline

10-FRAMES

100-CHART

1	2	3	4	5	6	7	8	9	10
11	12	13	14	15	16	17	18	19	20
21	22	23	24	25	26	27	28	29	30
31	32	33	34	35	36	37	38	39	40
41	42	43	44	45	46	47	48	49	50
51	52	53	54	55	56	57	58	59	60
61	62	63	64	65	66	67	68	69	70
71	72	73	74	75	76	77	78	79	80
81	82	83	84	85	86	87	88	89	90
91	92	93	94	95	96	97	98	99	100

NUMBER PATHS

NUMBER LINES

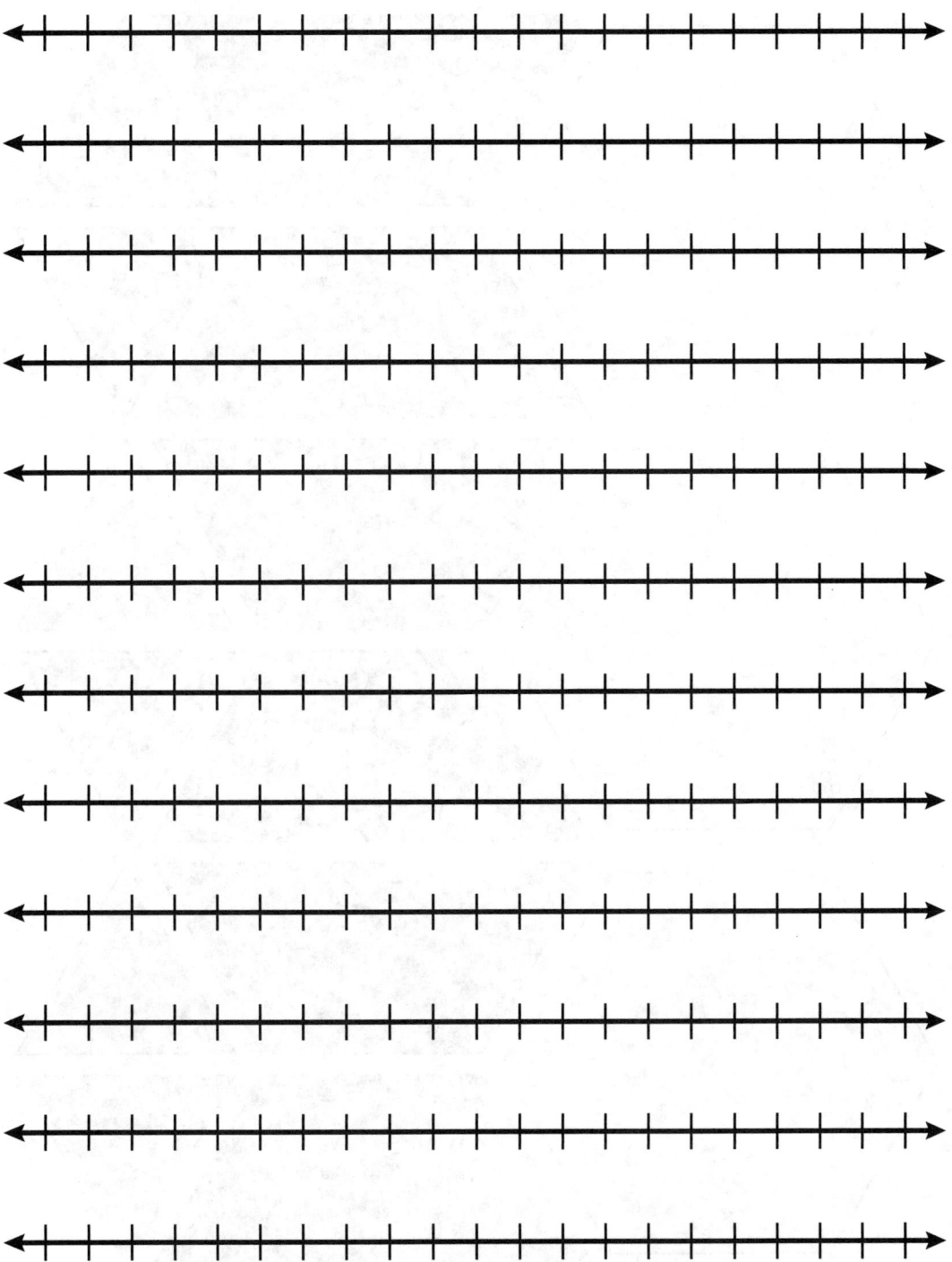

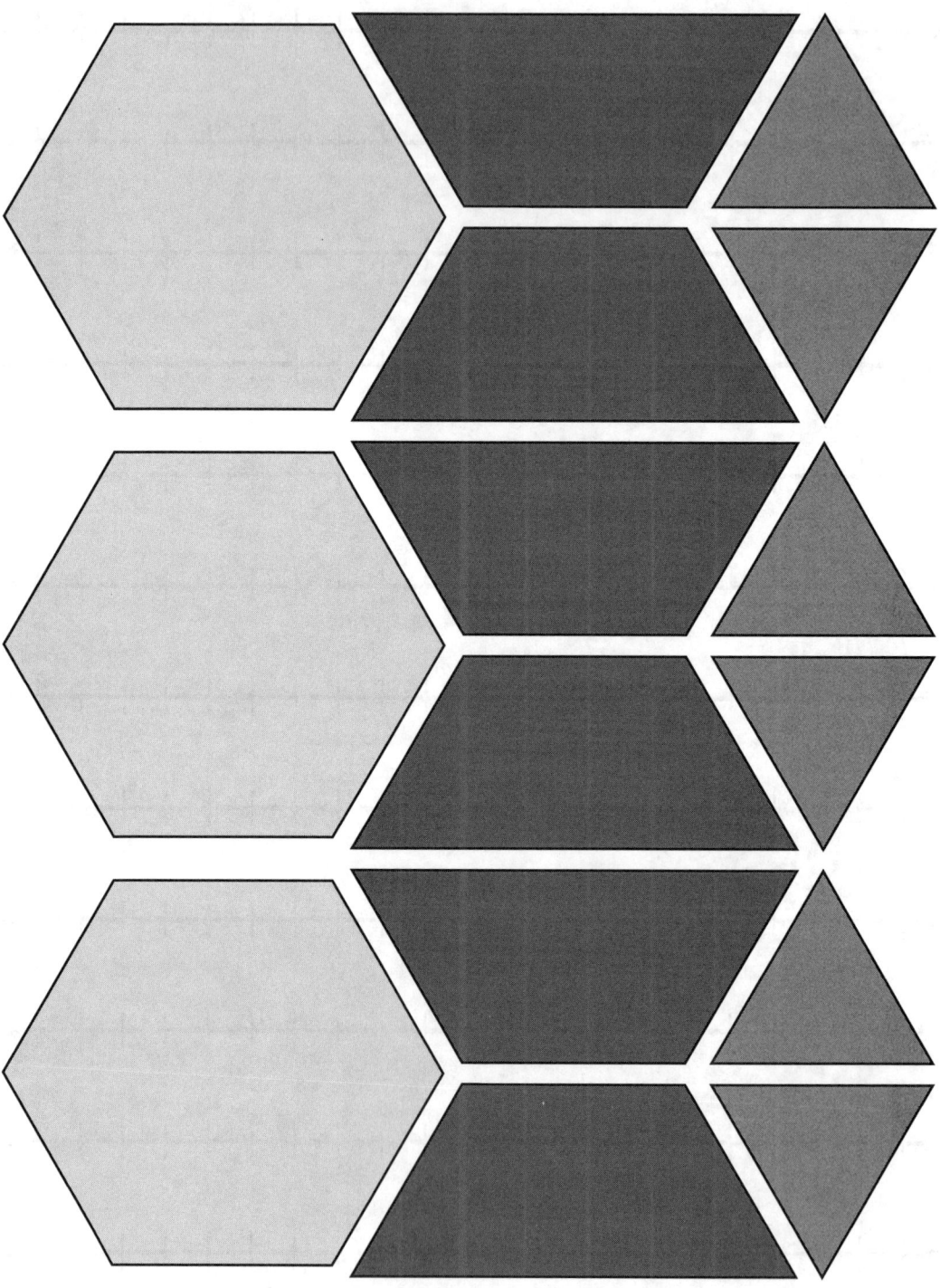

PATTERN BLOCKS (page 1)

PATTERN BLOCKS (page 2)

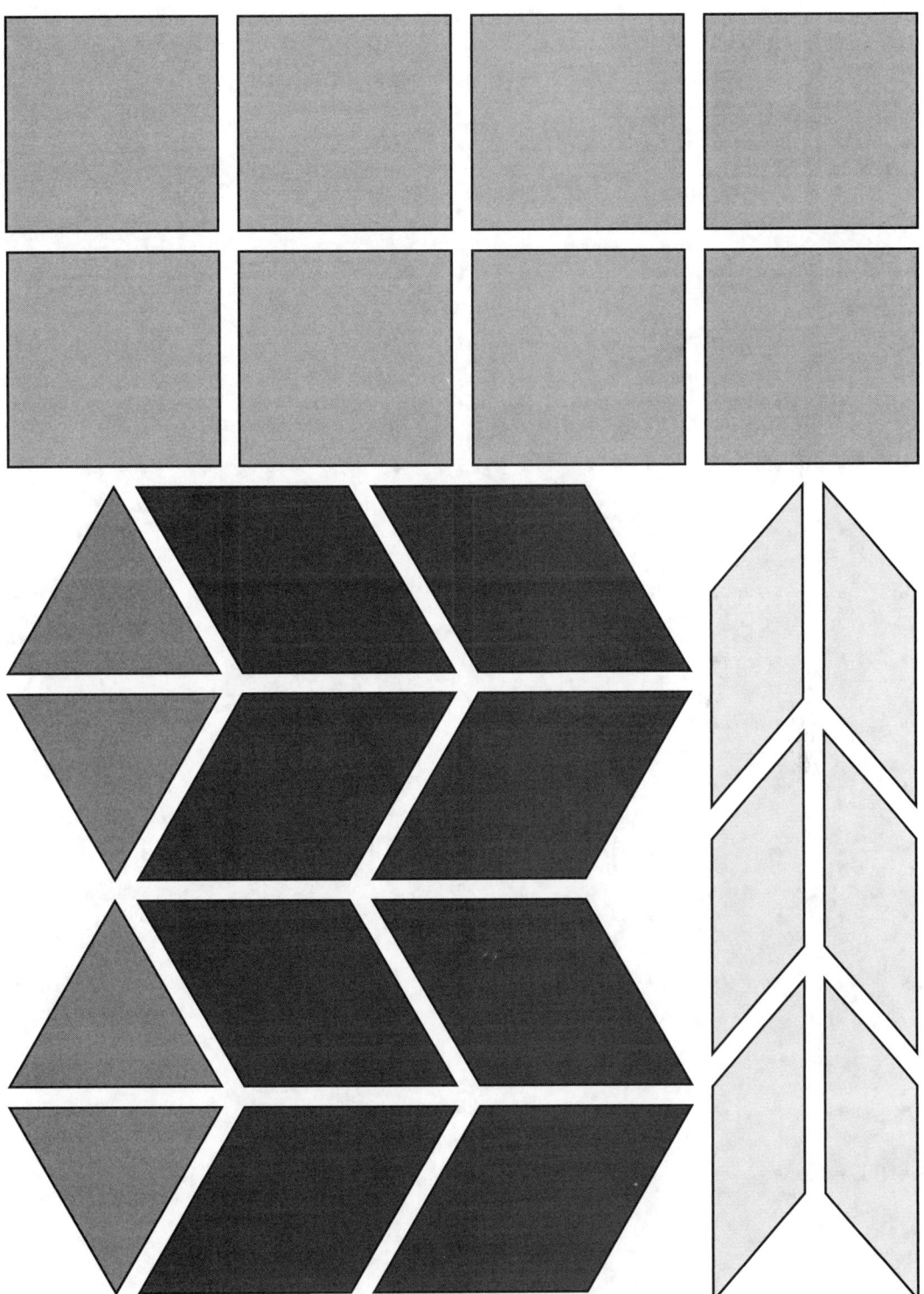

DOT PAPER

BASE TEN BLOCKS

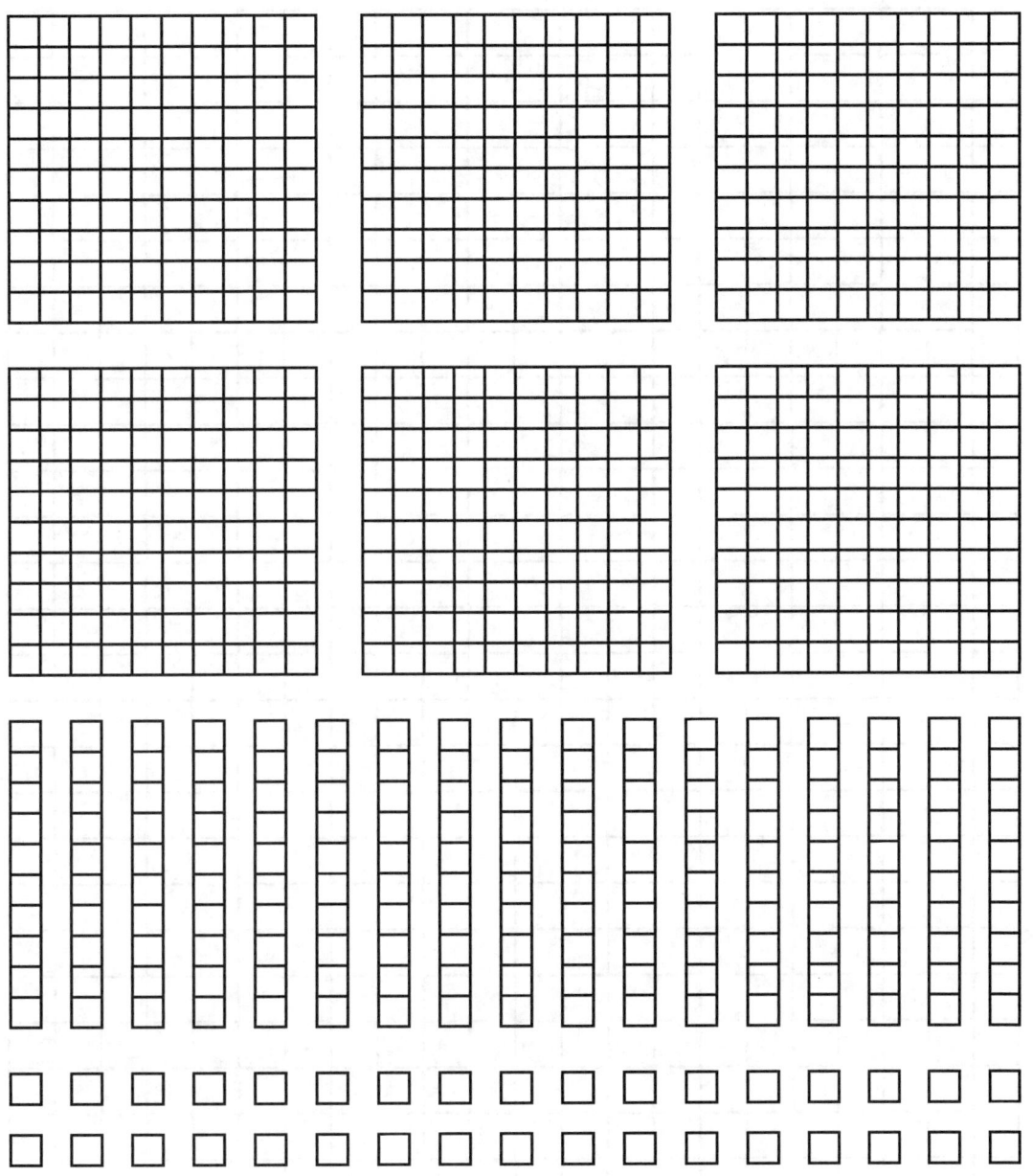

GRID PAPER

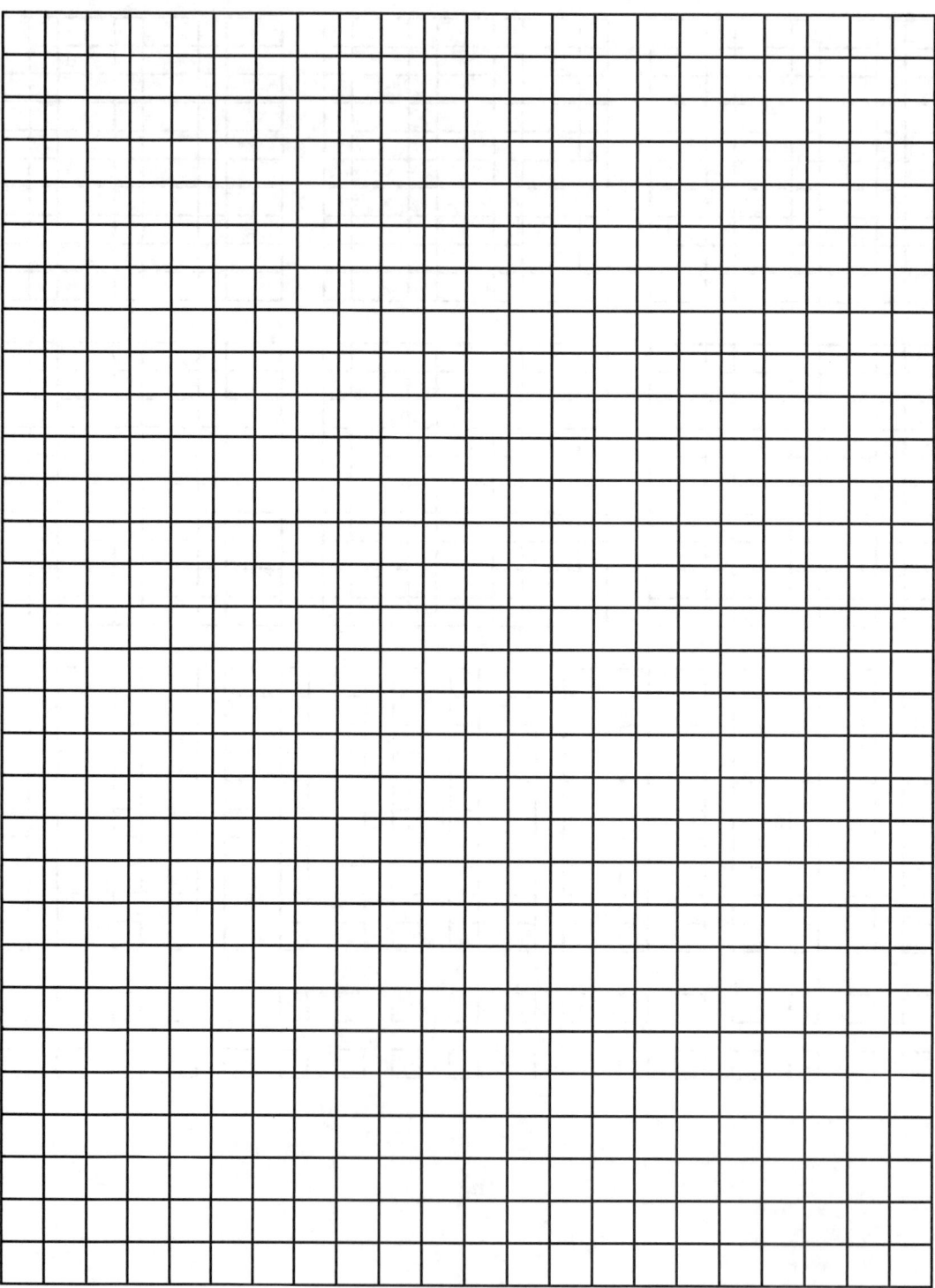

FRACTION STRIPS

1																			

1

$\frac{1}{2}$ $\quad\quad$ $\frac{1}{2}$

$\frac{1}{3}$ $\quad$ $\frac{1}{3}$ $\quad$ $\frac{1}{3}$

$\frac{1}{4}$ $\quad$ $\frac{1}{4}$ $\quad$ $\frac{1}{4}$ $\quad$ $\frac{1}{4}$

$\frac{1}{5}$ $\quad$ $\frac{1}{5}$ $\quad$ $\frac{1}{5}$ $\quad$ $\frac{1}{5}$ $\quad$ $\frac{1}{5}$

$\frac{1}{6}$ $\quad$ $\frac{1}{6}$ $\quad$ $\frac{1}{6}$ $\quad$ $\frac{1}{6}$ $\quad$ $\frac{1}{6}$ $\quad$ $\frac{1}{6}$

$\frac{1}{8}$ $\quad$ $\frac{1}{8}$ $\quad$ $\frac{1}{8}$ $\quad$ $\frac{1}{8}$ $\quad$ $\frac{1}{8}$ $\quad$ $\frac{1}{8}$ $\quad$ $\frac{1}{8}$ $\quad$ $\frac{1}{8}$

$\frac{1}{9}$ $\quad$ $\frac{1}{9}$ $\quad$ $\frac{1}{9}$ $\quad$ $\frac{1}{9}$ $\quad$ $\frac{1}{9}$ $\quad$ $\frac{1}{9}$ $\quad$ $\frac{1}{9}$ $\quad$ $\frac{1}{9}$ $\quad$ $\frac{1}{9}$

$\frac{1}{10}$ $\quad$ $\frac{1}{10}$ $\quad$ $\frac{1}{10}$ $\quad$ $\frac{1}{10}$ $\quad$ $\frac{1}{10}$ $\quad$ $\frac{1}{10}$ $\quad$ $\frac{1}{10}$ $\quad$ $\frac{1}{10}$ $\quad$ $\frac{1}{10}$ $\quad$ $\frac{1}{10}$

$\frac{1}{12}$ $\quad$ $\frac{1}{12}$ $\quad$ $\frac{1}{12}$ $\quad$ $\frac{1}{12}$ $\quad$ $\frac{1}{12}$ $\quad$ $\frac{1}{12}$ $\quad$ $\frac{1}{12}$ $\quad$ $\frac{1}{12}$ $\quad$ $\frac{1}{12}$ $\quad$ $\frac{1}{12}$ $\quad$ $\frac{1}{12}$ $\quad$ $\frac{1}{12}$

$\frac{1}{15}$ $\quad$ $\frac{1}{15}$ $\quad$ $\frac{1}{15}$ $\quad$ $\frac{1}{15}$ $\quad$ $\frac{1}{15}$ $\quad$ $\frac{1}{15}$ $\quad$ $\frac{1}{15}$ $\quad$ $\frac{1}{15}$ $\quad$ $\frac{1}{15}$ $\quad$ $\frac{1}{15}$ $\quad$ $\frac{1}{15}$ $\quad$ $\frac{1}{15}$ $\quad$ $\frac{1}{15}$ $\quad$ $\frac{1}{15}$ $\quad$ $\frac{1}{15}$

$\frac{1}{18}$ $\quad$ $\frac{1}{18}$ $\quad$ $\frac{1}{18}$ $\quad$ $\frac{1}{18}$ $\quad$ $\frac{1}{18}$ $\quad$ $\frac{1}{18}$ $\quad$ $\frac{1}{18}$ $\quad$ $\frac{1}{18}$ $\quad$ $\frac{1}{18}$ $\quad$ $\frac{1}{18}$ $\quad$ $\frac{1}{18}$ $\quad$ $\frac{1}{18}$ $\quad$ $\frac{1}{18}$ $\quad$ $\frac{1}{18}$ $\quad$ $\frac{1}{18}$ $\quad$ $\frac{1}{18}$ $\quad$ $\frac{1}{18}$ $\quad$ $\frac{1}{18}$

$\frac{1}{20}$ $\quad$ $\frac{1}{20}$ $\quad$ $\frac{1}{20}$ $\quad$ $\frac{1}{20}$ $\quad$ $\frac{1}{20}$ $\quad$ $\frac{1}{20}$ $\quad$ $\frac{1}{20}$ $\quad$ $\frac{1}{20}$ $\quad$ $\frac{1}{20}$ $\quad$ $\frac{1}{20}$ $\quad$ $\frac{1}{20}$ $\quad$ $\frac{1}{20}$ $\quad$ $\frac{1}{20}$ $\quad$ $\frac{1}{20}$ $\quad$ $\frac{1}{20}$ $\quad$ $\frac{1}{20}$ $\quad$ $\frac{1}{20}$ $\quad$ $\frac{1}{20}$ $\quad$ $\frac{1}{20}$ $\quad$ $\frac{1}{20}$

10 × 10 GRID

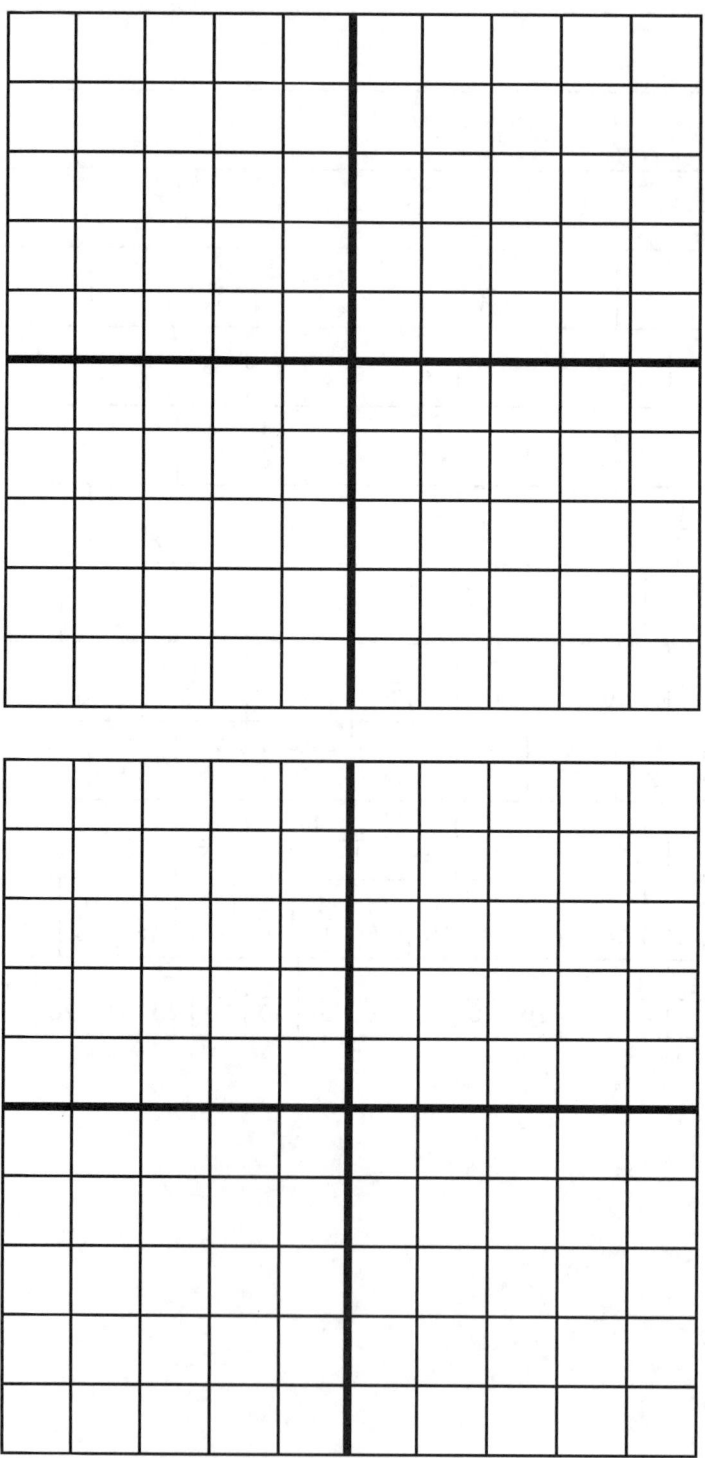

DOUBLE NUMBER LINES

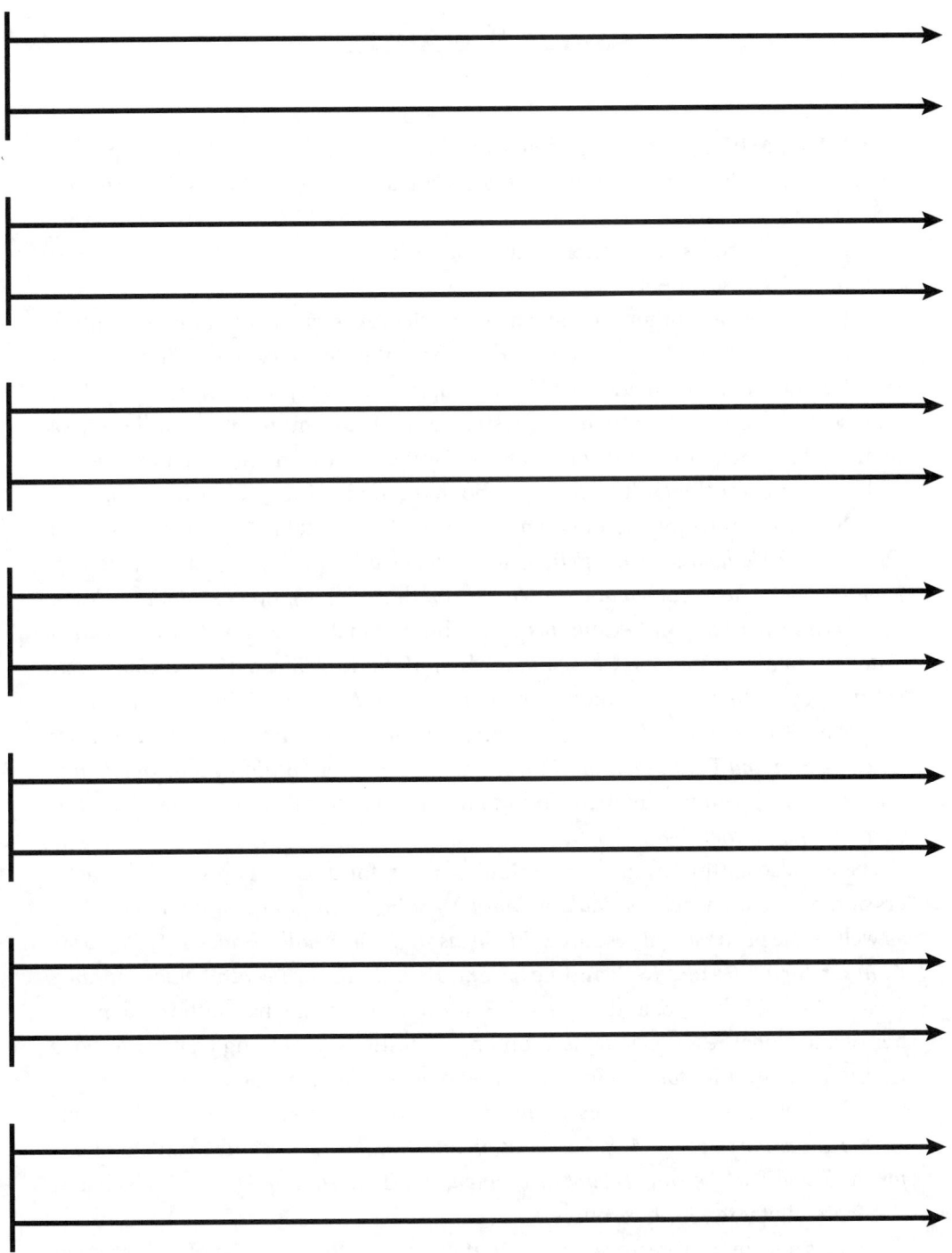

About the Author

MARIAN SMALL is a bestselling author, conference speaker, and trusted professional development consultant in the USA, Canada, and around the world. She has helped hundreds of schools and many thousands of K–12 teachers to grow more confident and successful in their mathematics instruction, feedback to students, and formative assessment.

She has been an author on many mathematics text series at both the elementary and the secondary levels. She has served on the author team for the National Council of Teachers of Mathematics (NCTM) Navigation series (pre-K–2), as the NCTM representative on the Mathcounts question writing committee for middle school mathematics competitions throughout the United States, and as a member of the editorial panel for the NCTM 2011 yearbook on motivation and disposition.

Dr. Small is probably best known for her books *Good Questions: Great Ways to Differentiate Mathematics Instruction,* and *More Good Questions: Great Ways to Differentiate Secondary Mathematics Instruction* (with Amy Lin). *Eyes on Math: A Visual Approach to Teaching Math Concepts* was published in 2013, as was *Uncomplicating Fractions to Meet Common Core Standards in Math, K–7.* She authored *Uncomplicating Algebra to Meet Common Core Standards in Math, K–8,* in 2014, *Teaching Mathematical Thinking: Tasks and Questions to Strengthen Practices and Processes* in 2017, *Fun and Fundamental Math for Young Children: Building a Strong Foundation in PreK–Grade 2* in 2018, and *Math That Matters: Targeted Assessment and Feedback for Grades 3–8* in 2019.

She is also author of the four editions of a text for university preservice teachers and practicing teachers, *Making Math Meaningful to Canadian Students: K–8,* as well as the professional resources *Big Ideas from Dr. Small: Grades 4–8*; *Big Ideas from Dr. Small: Grades K–3*; and *Leaps and Bounds Toward Math Understanding: Grades 3–4, Grades 5–6,* and *Grades 7–8.* More recently, she has authored a number of additional resources focused on open questions, including *Open Questions for the Three-Part Lesson, K–3: Numeration and Number Sense*; *Open Questions for the Three-Part Lesson, K–3: Measurement and Pattern and Algebra*; *Open Questions for the Three-Part Lesson, 4–8: Numeration and Number Sense*; and *Open Questions for the Three-Part Lesson, 4–8: Measurement and Pattern and Algebra,* as well as a K–8 core digital teaching resource.

She is a former professor of mathematics and the former dean of education at the University of New Brunswick.